ENGINEERING WITH LEGO® BRICKS AND ROBOLAB™

The official guide to ROBOLAB

Eric L. Wang
Department of Mechanical Engineering
University of Nevada, Reno

In consultation with
Chris Rogers
Department of Mechanical Engineering
Tufts University

College House Enterprises, LLC
Knoxville, Tennessee

Version 0.6: September 2003
Second Printing: April 2004

Copyright © 2003 by Eric L. Wang

Reproduction or translation of any part of this work beyond that permitted by Sections 107 and 108 of the 1976 United States Copyright Act without permission of the copyright owner is unlawful. Requests or further information should be addressed to Dr. Eric Wang, University of Nevada, Department of Mechanical Engineering, Mail stop 312, Reno, NV 89557. email: eric.wang@unr.edu.

This material is based upon work supported by the National Science Foundation under Grant No. 9950741 Any opinions, findings and conclusions or recommendations expressed in this material are those of the author(s) and do not necessarily reflect the views of the National Science Foundation.

This textbook is intended to provide accurate and authoritative information regarding the topics described. It is distributed and sold with the understanding that the publisher is not engaged in providing legal, accounting, engineering or other professional services. If legal advice or other expertise advice is required, the services of a recognized professional should be retained.

The manuscript was prepared using Word 2000 with 11 point Times New Roman font. Publishing and Printing Inc., Knoxville, TN printed this workbook.

College House Enterprises, LLC.
5713 Glen Cove Drive
Knoxville, TN 37919, U. S. A.
Phone   (865) 558 6111
FAX   (865) 584 1766
e mail   jdally@www.collegehousebooks.com
Visit our web site   http://www.collegehousebooks.com

ISBN 0-9700675-1-8

PREFACE: THE WORKBOOK VERSION

About the workbook:

This first draft represents a work in progress. The final book will include 8 chapters along with several appendices and on-line reference material. Unfortunately, the completion of the full text is not expected until sometime in the year 2004.

Many people have asked that we publish something while the full text is being completed. To meet this demand, we have produced the first 4 chapters of the book in workbook form, which you are now holding in your hands.

About LEGO Bricks and ROBOLAB:

I strongly feel that engineering education can, and should be, fun. So when I first heard about using LEGO bricks to teach engineering, I was instantly intrigued. I had spent way too many hours playing with LEGO bricks when I was a kid and this sounded like the perfect excuse to start playing with them again. After a couple of terrible initial experiences with the standard Mindstorms programming language and NQC, I hit upon ROBOLAB. It worked great, mainly due to the very short learning curve. The only problem was there wasn't much documentation beyond the Getting Started teacher's manuals.

Thus, the idea for a book was hatched. Chris Rogers, the creator of ROBOLAB, came to Reno for a conference and over dinner we hashed out the skill badge concept and the outline for this book. We were both pretty adamant that the book focus on project based learning. Our experience has shown that hands-on engineering is one of the best ways to learn. Plus, it tends to break down the barriers between the instructor and the student.

With that said, this workbook is intended to be used by first year engineering students, although others may find it is useful as a ROBOLAB reference guide. We've tried to think up design challenges that have no ceiling. You can accomplish the tasks with a moderate amount of effort or you can make it as complicated as you want.

I want to see how far I can push LEGO bricks and ROBOLAB, both of which were developed for children – not engineering students. I can't speak for Chris, but I think he's of similar mind. We've had to restrain ourselves quite a bit because we keep devising (somewhat diabolically) ever more complex challenges. Fortunately for our students, most of these I have reserved for the last four chapters. On the other hand, if you are somewhat deranged like me, you will just have to wait until 2004 when the full text is released.

Eric Wang
Reno, NV
August 2002

TABLE OF CONTENTS

CHAPTER 1 1-1
- 1.1 ORGANIZATION OF THIS BOOK 1-1
- 1.2 LEGO MINDSTORMS™ HARDWARE 1-3
  - 1.2.1 The Programmable Brick: the RCX 1-4
  - 1.2.2 Output and Input Devices 1-5
- 1.3 ROBOLAB SOFTWARE 1-7
  - 1.3.1 Administrator 1-8
  - 1.3.2 Programmer 1-9
  - 1.3.3 Investigator 1-10
  - 1.3.4 ROBOLAB Version 2.5.2 1-10
- 1.4 DESIGN SKILLS 1-10
  - 1.4.1 General LEGO Building Tips 1-11
  - 1.4.2 Gears and Axles 1-13
  - 1.4.3 Getting Around on Wheels 1-17
  - 1.4.4 Getting Around on Legs 1-22
  - 1.4.5 Bumpers and Sensors 1-25
  - 1.4.6 Grippers and Claws 1-29
  - 1.4.7 Creativity & Aesthetics 1-31
- 1.5 ONLINE RESOURCES 1-33
- 1.6 DESIGN CHALLENGES 1-35
  - 1.6.1 Team Communication 1-36
  - 1.6.2 Drag Race 1-37
  - 1.6.3 South Pointing Chariot 1-38
  - 1.6.4 Heavy Lifting 1-39
  - 1.6.5 Crash Test Dummy 1-40
  - 1.6.6 Semester Clock 1-41

CHAPTER 2 2-1
- 2.1 GREEN CHALLENGES 2-2
  - 2.1.1 The Steepest Incline 2-2
  - 2.1.2 Tug-of-War 2-3
  - 2.1.3 Going the Distance 2-4
  - 2.1.4 Tunnel Vision 2-6
  - 2.1.5 Wallace & Gromit™ 2-7
  - 2.1.6 Line Follower 2-8
  - 2.1.7 Speed Walking 2-9
  - 2.1.8 How Fast is That? 2-10
- 2.2 PILOT BASICS 2-11
- 2.3 THE BASIC PILOT BADGE 2-12
  - 2.3.1 Outputs 2-13
  - 2.3.2 The "Wait For" Functions 2-14
  - 2.3.3 Modifiers 2-15
  - 2.3.4 Sample Pilot Level 4 Programs 2-16
  - 2.3.5 Notes About Using the LEGO Light Sensor 2-16
  - 2.3.6 Steps, Run Mode, Printing, and Saving 2-19

2.4	RELATION TO TEXT-BASED PROGRAMMING	2-21
2.5	LIMITATIONS OF PILOT PROGRAMMING	2-23
2.6	INVENTOR BASICS	2-24
2.6.1	The Functions Palette	2-25
2.6.2	The Tools Palette	2-26
2.6.3	Getting Help	2-29
2.7	THE BASIC INVENTOR BADGE	2-30
2.7.1	Outputs	2-30
2.7.2	The "Wait For" Functions	2-32
2.7.3	Modifiers	2-35
2.7.4	Getting to the Source	2-36
2.7.5	Sample Programs	2-38
2.7.6	What's Next?	2-39

CHAPTER 3..3-1

3.1	WHITE CHALLENGES	3-2
3.1.1	Wall Follower	3-2
3.1.2	Remote Control	3-4
3.1.3	LEGO Brick Recycler	3-5
3.1.4	Stay Inside the Box	3-6
3.1.5	Fetch the Light	3-7
3.1.6	Robot Zoo	3-8
3.1.7	Edge Detector (a.k.a Barcode Reader)	3-9
3.1.8	Round and Round (a.k.a. shaft encoder)	3-11
3.1.9	Two steps forward, one step back	3-13
3.1.10	There and back again	3-14
3.1.11	Swinging with Gravity	3-15
3.1.12	Tomb Raider	3-16
3.2	THE STRUCTURES BADGE	3-17
3.2.1	Jumps	3-17
3.2.2	Loops	3-19
3.2.3	Forks	3-21
3.3	THE CONTAINERS BADGE	3-24
3.3.1	Container basics	3-24
3.3.2	Container Sub-menu	3-26
3.3.3	Container Wait For, Loop, and Fork functions	3-27
3.3.4	Integer math	3-28
3.3.5	Generic Container	3-28
3.3.6	Container Examples	3-29
3.4	THE TASKS BADGE	3-30
3.4.1	Tasks Splits	3-30
3.4.2	Starting and Stopping Tasks	3-32
3.5	INVESTIGATOR BASICS	3-34
3.6	THE BASIC INVESTIGATOR BADGE	3-36
3.6.1	Program Area	3-36
3.6.2	Upload Area	3-40
3.6.3	View and Compare Area	3-44
3.6.4	Compute Area	3-46
3.6.5	Journal Area	3-50

CHAPTER 4 4-1

- 4.1 BLACK CHALLENGES 4-2
  - 4.1.1 LEGO Slot Machine 4-2
  - 4.1.2 The Sound of Music 4-4
  - 4.1.3 Bionic Bat 4-5
  - 4.1.4 Speed Walking (revisited) 4-6
  - 4.1.5 Simon Says 4-7
  - 4.1.6 Demolition Derby 4-9
  - 4.1.7 Animal Behavior 4-10
  - 4.1.8 Can You hear Me Now? 4-12
  - 4.1.9 CodeMaster 4-13
  - 4.1.10 Reverse Engineering 4-15
  - 4.1.11 Stay Away From the Light 4-16
- 4.2 THE EVENTS BADGE 4-18
  - 4.2.1 What's an event? 4-18
  - 4.2.2 How to program an event 4-19
  - 4.2.3 Stopping and re-starting event monitoring 4-20
  - 4.2.4 Multiple events 4-21
  - 4.2.5 Events and tasks 4-22
  - 4.2.6 Event Examples 4-23
- 4.3 THE MUSIC BADGE 4-24
  - 4.3.1 basic Notes 4-25
  - 4.3.2 Music Scrolls 4-26
- 4.4 THE RCX COMMUNICATION BADGE 4-27
  - 4.4.1 Mail 4-27
  - 4.4.2 Set Display 4-29
  - 4.4.3 RCX Communication Examples 4-30
- 4.5 THE DIRECT MODE BADGE 4-31
  - 4.5.1 Begin and End Direct Mode 4-31
  - 4.5.2 Wait for RCX to be in view 4-32
  - 4.5.3 Read Value 4-33
  - 4.5.4 RCX Battery Power 4-34
  - 4.5.5 No Mail in Direct Mode 4-35
  - 4.5.6 Running Direct Mode on top of Remote Mode 4-36
- 4.6 THE INTERNET BADGE 4-37
  - 4.6.1 Nomenclature: Remote, Direct Internet, and Local 4-38
  - 4.6.2 Configuring the ROBOLAB Internet Server 4-39
  - 4.6.3 Internet Begin and End 4-39
  - 4.6.4 Internet Direct Mode 4-41
  - 4.6.5 Controlling Two RCX's Simultaneously 4-42
- 4.7 THE ADVANCED OUTPUT BADGE 4-46
  - 4.7.1 Advanced Output sub-menu 4-47
  - 4.7.2 Pulse Width Modulation 4-47
  - 4.7.3 Controlling Motor Power 4-48
  - 4.7.4 Power versus Speed 4-49

CHAPTER 1

LEGO & ROBOLAB Basics

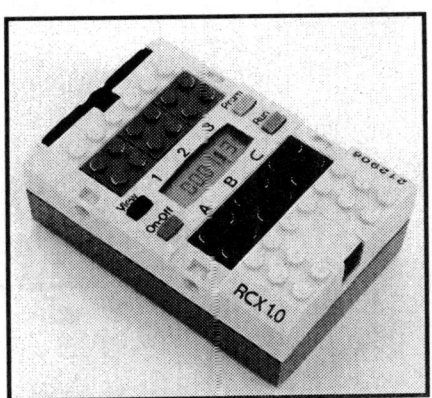

If you are like most people, you probably have not played with LEGO bricks since you were a child. However LEGO bricks are wonderful for learning about many physics, engineering, and computer science principles and skills.

1.1 Organization of This Book

Rather than the traditional homework problem set at the end of each chapter, this book utilizes **design challenges.** In a design challenge we define a task that must be completed. It is your job to design, build, and program a robot to accomplish this task. We purposely, and somewhat diabolically, place very few constraints (often called 'rules') on possible solutions so that you have plenty of room to stretch your creative legs. Most, if not all, of the challenges have no single solution.

The real world if full of design challenges and their solutions. Just take a trip down to your local department store and you will find dozens of devices that accomplish the simple task of "toasting a piece of bread."

You and your instructor will monitor progress in your learning by using a **skill badge** concept. Skill badges are earned for successfully completing a challenge. The skill badges required are indicated at the start of each challenge. How difficult a particular challenge is will depend on how many of the required skill badges you already have earned. In general, you should not try to acquire more than three new skill badges with any single challenge (unless you have an extremely good teaching assistant or have a lot of time on your hands).

The challenges are organized into 5 levels of difficulty: Green (easiest), White, black, Blue and Red (most difficult). At each level we assume you have earned all the skill badges from the easier levels. With this said, however, we are aware that it is possible to skip levels depending on the skill badges sought.

Unlike most textbooks, you will not read this book sequentially. That is, we fully expect that you will skip back and forth between sections and chapters as you acquire the various skill badges. You will also find that each chapter is organized backwards from a standard textbook; the challenges appear at the beginning of the chapter followed by the sections aimed at teaching a particular skill. In organizing the book in this way, we are encouraging you to develop **self learning skills**, which are crucial to your future success as an Engineer.

The remainder of Chapter 1 deals with the nuts, bolts and electrons of the LEGO Mindstorms hardware and the ROBOLAB software. We presume that you have already successfully installed the batteries, installed the software, downloaded the firmware, and watched the video included on the CD-ROM. If you have not done so or are having trouble, please consult Appendix C for troubleshooting tips.

Chapters 2 through 6 each cover one challenge level and have several challenges at the beginning of each chapter. Figure 1.1 shows the layout of this book and the skill badges associated with each challenge level. There are a total of 28 skills badges that can be earned. Most introduction to engineering courses will cover about 10-12 skills badges.

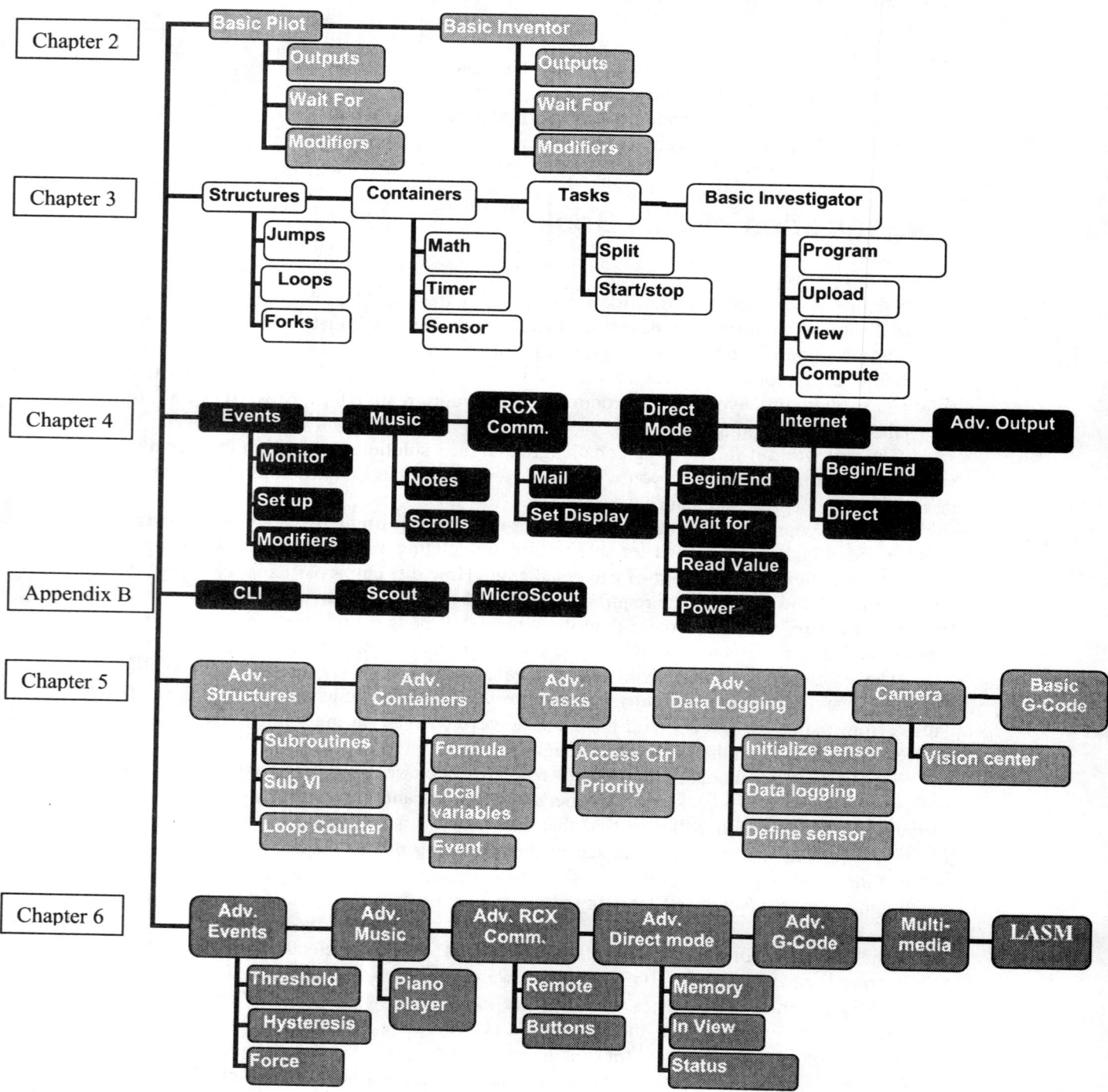

Figure 1.1. Organization of this book showing the challenge levels and skill badges.

1.2 LEGO Mindstorms™ Hardware

LEGO Mindstorms™ is a relatively new brand of the LEGO Group. It came about in 1998 after being inspired by the book *Mindstorms* by Dr. Seymour Papert at the Massachusetts Institute of Technology's Media Lab.

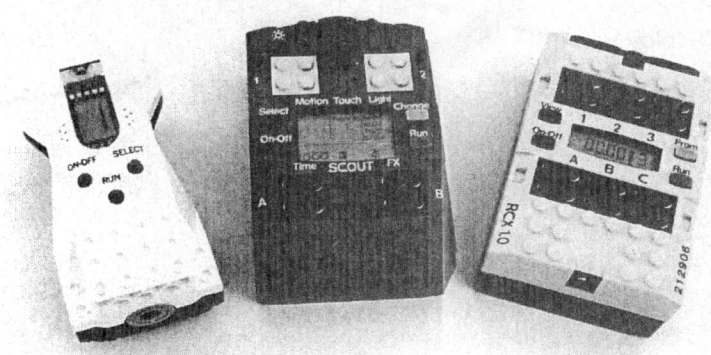

Figure 1.2. LEGO smart bricks: the Microscout (left), Scout, and RCX (right).

The Mindstorms family includes three "smart bricks:" the RCX, the Scout, and the Microscout. All three smart bricks contain microprocessors, inputs, and outputs which allow you to create moble robots that can react to their environment. The RCX evolved from the crickets and handyboard developed at MIT. Fred Martin's book, *Robotic Explorations*, provides an excellent history of the MIT programmable brick.

The Scout is based on the RCX but was developed as a platform which does not require a computer to program (although ROBOLAB can be used to program it – see Appendix B). At the time of this writing LEGO had recently discontinued the Scout, but it can still be purchased as some retail stores and Internet auction sites.

The Microscout is the third smart brick developed by LEGO. The Microscout can either execute one of seven internal programs or can be programmed via a visible light link (VLL) on a PC or Scout brick. Strangely enough, the Microscout cannot be directly programmed by the RCX, only the Scout brick can do this.

All three smart bricks can be programmed with a wide variety of software programs such as pbforth, NQC, LeJOS, BrickCommand, and BrickOS but this book will concentrate on ROBOLAB as the programming environment. This book will also concentrate on only the RCX. Information and design challenges for the Scout and Microscout can be found in Appendix B.

1.2.1 The Programmable Brick: the RCX

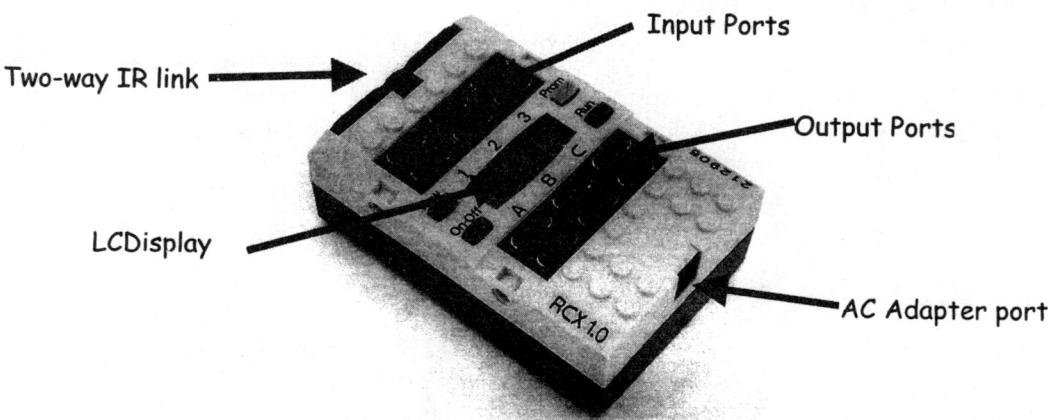

Figure 1.3
The RCX programmable brick.

The Robotic Command Explorer, or RCX, is the brain of the LEGO Mindstorms kit. There are six ports on the RCX. The three upper ports (1, 2, and 3) are the input ports; connect sensors to these. The three lower ports (A, B, and C) are the output ports; connect motors, lights, and other output devices to these.

The RCX also comes equipped with a LCD screen for displaying useful information, four buttons for activating the RCX, an internal speaker for playing sounds, and an infrared (IR) communications port. Retail versions of the RCX do not come with the AC adapter port, but the version available from the Pitsco LEGO Educational Division is equipped with the port. We highly recommend purchasing the RCX version 1.0 (W979709) that is available online at http://www.pldstore.com/. While an AC adapter cord limits the mobility of the robots built, it makes sensor data and motor output much more reliable and is more environmentally friendly than disposable batteries.

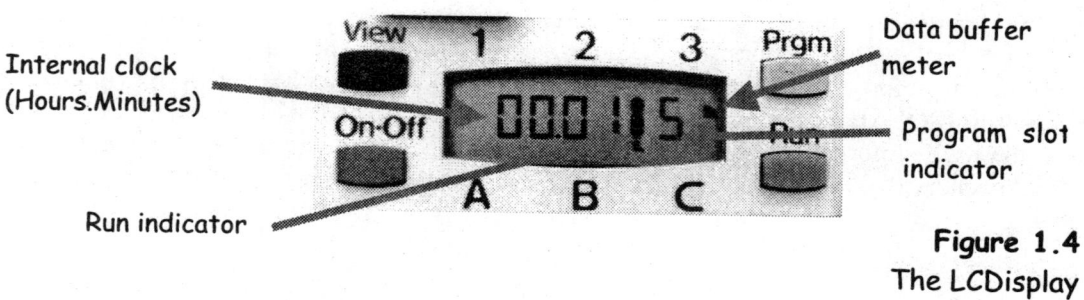

Figure 1.4
The LCDisplay

The RCX display normally displays the internal clock in hour.minutes since it was last turned on. Turning the RCX on and off resets the internal clock. When a program is running the run indicator is animated. The Program slot indicator displays which of the 5 program slots is currently selected. There is also an icon that indicates how full the internal data buffer is (yes, the RCX can function as a data acquisition device to collect and store data).

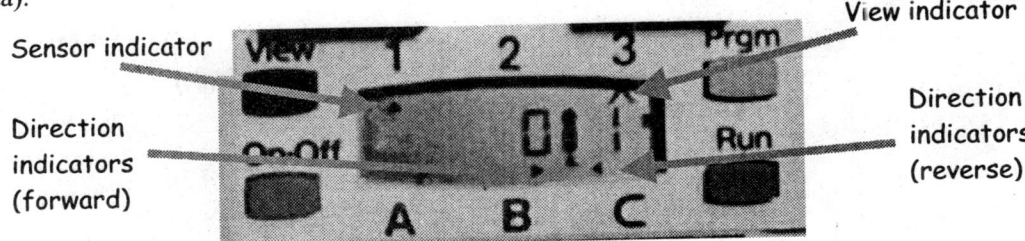

Figure 1.5
LCD indicator symbols

When a program is running, the LCD also shows which sensors are active (sensor indicator) and which direction the motor output is running (direction indicator). By depressing the **View** (black) button, you can also monitor the current sensor reading on any of the 3 input ports or power levels on the 3 output ports.

> Pressing the **View** and **Run** buttons simultaneously turns on output A, B, or C in the forward direction. Likewise, **View + Prgm** will turn A, B, or C in the reverse direction. This can be useful for determining which direction a motor will spin when connected to the output port.

1.2.2 Output and Input Devices

The RCX is also outfitted with an internal speaker for playing sounds and music (in fact there are 2 music skill badges). External outputs are connected to the output ports A, B, and C. In addition to the motors that come with the Mindstorms kit there are several other LEGO output devices. All of the output devices shown in Table 1.1 can be attached to any of the output ports (A, B, or C). As will be discussed in Chapter 2, unlike motors the sound and light elements are not affected by the orientation of the lead wires.

The Mindstorms kit comes with two touch and one light sensor, but there are several other LEGO sensors commercially available as shown in Table 1.2. All of the input devices can be attached to any of the input ports (1, 2, or 3). The orientation of the lead wires does not matter.

Additionally, there are dozens of input and output devices that you can make yourself if you have some basic electronics skills. These are beyond the scope of this text, but there are several good books and websites [e.g. *Robotic Explorations* by Martin, *The Unofficial Guide to LEGO MINDSTORMS Robots* by Knudsen]. The ROBOLAB Reference Guide also has lot of information on defining your own sensors. It's located in the ROBOLAB directory under the Support Material folder in ROBOLAB versions 2.5.1 and higher. There are also several good websites (see section 1.6) that discuss how to connect more than one sensor to a single input port, however, this too is beyond the scope of this text.

Table 1.1. Output Devices

Output Devices		Notes
	Motor (internally geared)	Reversing the orientation of the lead wire on either the RCX or the motor will reverse the direction of rotation of the motor.
	Micromotor	Reversing the orientation of the lead wire on either the RCX or the motor will reverse the direction of rotation of the motor.
	9V Motor	Reversing the orientation of the lead wire on either the RCX or the motor will reverse the direction of rotation of the motor.
	Sound Element	Turning the top of the sound element 90 degrees changes the sound produced. Can be connected directly to the output port without a lead wire.
	Lamp Element	Can be attached to the top or bottom of the lead wire. Can be connected directly to the output port without a lead wire. Can also be stacked or placed side-by-side, with two or more elements on one output.

Table 1.2 Input Devices

Sensor		Description
	Touch	The wire lead is attached to the front four studs on the top of the touch sensor. The orientation of the lead wire does not matter. Two or more touch sensors can be attached to a single input port; pushing any of them will register as a touch.
	CLI touch	The CLI touch sensor is similar to the standard touch sensor, except the lead wire is integrated into the device.
	Light	The light sensor records values between 1 (dark) and 100 (bright). The units of measurement do NOT correspond to any standard unit of light measurement.
	Rotation	The rotation sensor measures in increments of $1/16^{ths}$ of a rotation. Rotations can be positive or negative depending on which direction the sensor is rotated.
	Temperature	The temperature sensor can measure temperatures between -20 and $+50$ Celsius (-4 to 122°F). The brick and wire portions of the sensor should not get wet.
	DCP sensors	Many other sensors can also be used with the RCX with the use of a DCP sensor adapter. DCP sensors include: voltage, humidity, pH, current, temperature, sound, motion, and pressure.
	Camera	The camera is the only sensor that is not connected directly to the RCX. The RCX receives the camera sensor readings via the IR port from the computer using Direct Mode.

1.3 ROBOLAB Software

ROBOLAB is a graphical programming environment that is based on LabVIEW, available from National Instruments. This section provides a brief overview of the ROBOLAB software. Chapter 2 will cover programming techniques in detail.
ROBOLAB has three different graphical programming modes within it

Pilot is the easiest programming mode. It utilizes a serial programming environment to ensure that the program will always compile and execute.

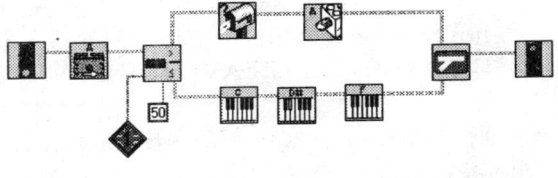

Inventor is the second programming mode. Program icons are "wired" together to create a program. Programs created can contain all the typical programming elements such as constants, variables, loops and functions.

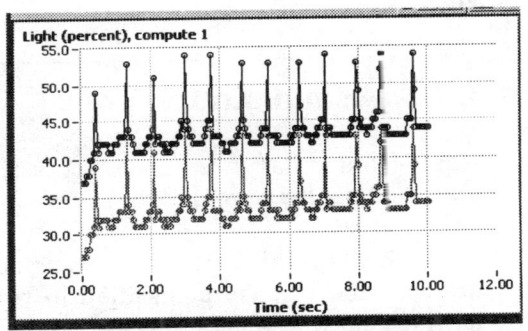

Investigator is the final programming mode. All of the features in Inventor mode are included along with the added feature of data logging and advanced data analysis.

Figure 1.6. ROBOLAB programming modes: Pilot (top), inventor, and investigator (bottom).

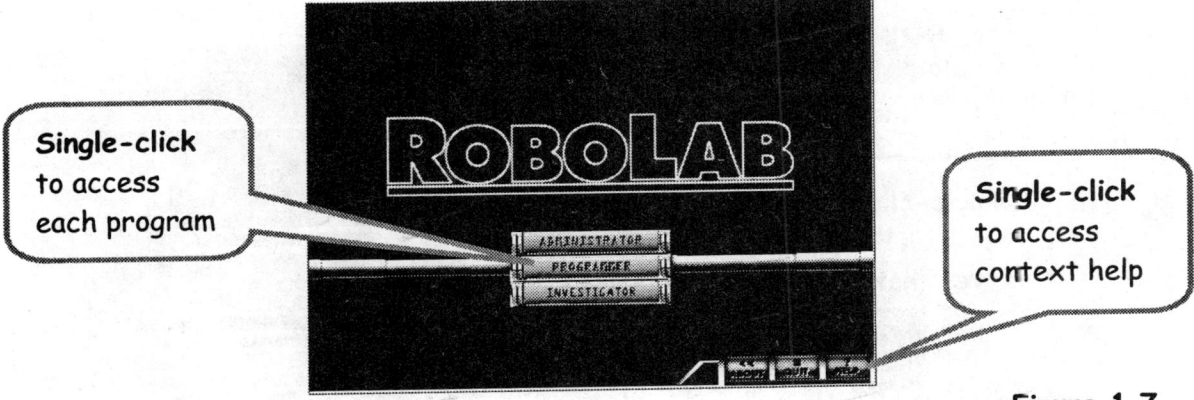

Single-click to access each program

Single-click to access context help

Figure 1.7 ROBOLAB introductory screen.

The introductory screen appears each time ROBOLAB is started. From this screen you can access the three main program areas:

- **Administrator:** this lets you adjust RCX settings, set default file locations, and download firmware.
- **Programmer:** this accesses Pilot and Inventor level programming modes. Programmer allows you to create robots that can interact with their environment, but it does not make use of the data acquisition capabilities of the RCX.
- **Investigator:** this accesses the Investigator level programming mode. Investigator level focuses on the data acquisition capabilities of the RCX. Additionally, through Investigator you can gain access to data processing commands.

The help button will open the Context Help window, which provides useful information on the object the cursor is over. WE STRONGLY ENCOURAGE ALL BEGINNERS TO LEAVE THE CONTEXT HELP TURNED ON AT ALL TIMES.

> **TIP:** If you do not see the Administrator button, hit function key **F5** to make it visible.

1.3.1 Administrator

Administrator has three areas accessed via the tabs at the bottom of the screen.

The **Administrator** tab allows you to setup and test the connection to the RCX and download the firmware.

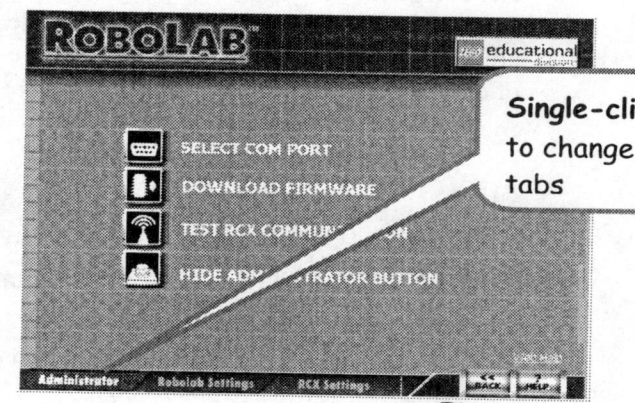

Figure 1.8
Administrator tab.

The **ROBOLAB Settings** tab allows you to add or delete new themes (program groups) and to set the default program paths.

Single-click to toggle between Programmer and Investigator settings

Figure 1.9
ROBOLAB settings tab.

In ROBOLAB version 2.5.0 and earlier, user programs must be stored in the Program Vault directory in order to be opened from within ROBOLAB. If you prefer to store your programs elsewhere (such as a floppy disk or c:\my documents) make sure to change the Program Vault directory.

The **RCX Settings** tab allows you to change several default RCX settings and poll the RCX for current battery level and firmware version. There are three settings which can be adjusted: the IR power setting, the lock on programs 1 and 2, and the powerdown time.

- **IR power setting**: the low setting will not broadcast as far as the high setting, but will save battery power. Setting this to high may be useful for RCX to RCX communication, but will consume more battery power.
- **Program 1 and 2 locks**: program slots 1 and 2 by default are locked, meaning you cannot overwrite them. This setting can be used to unlock the program slots 1 and 2, which can be useful if you are trying to store more than 3 custom programs. This setting can also be used to lock one of your custom programs by first unlocking slots 1 and 2, downloading your program to one of these slots, and then locking them again. Good for safeguarding a program you know works!
- **Powerdown time**: this sets the amount of inactive time the RCX will wait before turning itself off. The default time is 15 minutes. You may want to adjust this to a lower setting to conserve battery power.

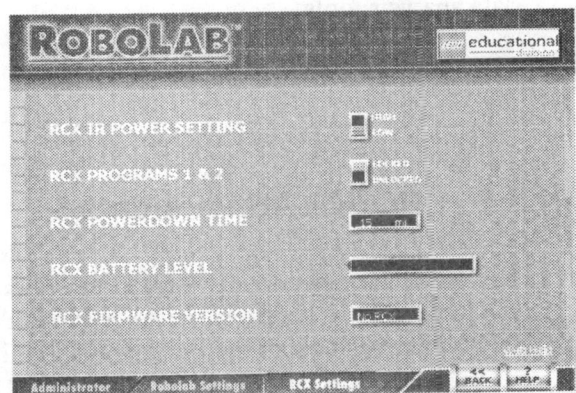

Figure 1.10
RCX Settings tab.

If you try to download a program to slot 1 or 2 while they are locked, the program will be downloaded to slot 3, overwriting any program in slot 3 in the process.

1.3.2 Programmer

From the Programmer window you can access both the Pilot and Inventor programming modes.

Pilot mode uses program templates to create programs that always compile, but are limited in complexity.

In **Inventor** mode you use a graphical programming interface to create programs limited only by your imagination.

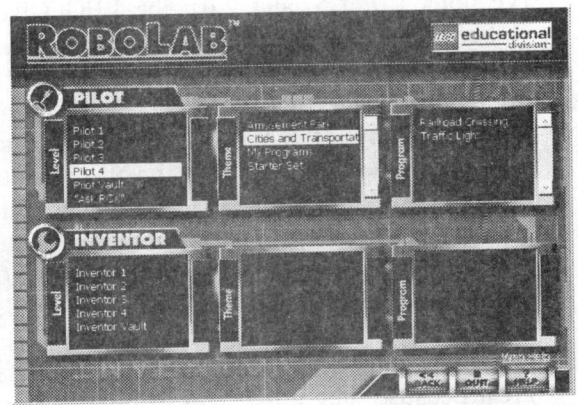

Figure 1.11
The Programmer main window.

1.3.3 Investigator

In Investigator mode, you can create programs that make use of the data acquisition capabilities of the RCX. You also have access to advanced programming and data analysis tools.

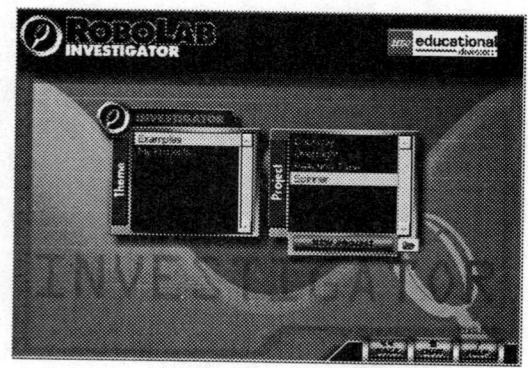

Figure 1.12
The Investigator main window.

1.3.4 ROBOLAB Version 2.5.2

This book was specifically written for the latest version of ROBOLAB, which is currently version 2.5.2. So naturally we are assuming that's the version you have installed. With that said, however, most of the information in this book can be used for earlier versions as well.

ROBOLAB versions 1.0 and 1.5 started it all, but only had Pilot and Inventor programming modes. In version 2.0 Investigator mode was added to exploit the data acquisition capabilities of the RCX.

More recently, version 2.5 was created to take full advantage of the new RCX firmware. It added camera support and advanced control. That brings us up to version 2.5.2 which fixed some minor bugs in version 2.5. If you are running earlier versions, you can download patches (bug fixes) for ROBOLAB at http://www.ceeo.tufts.edu/robolabatceeo/.

1.4 Design Skills

You may have played with LEGO bricks as a child, but if you are like most of our students you are bound to find out that contemporary LEGO bricks are not quite like the ones you used to play with. There are lots of strange looking pieces and a myriad of ways to put them together, some better than others.

In this section, we've compiled a series of hints and tips on how to build with the LEGO Mindstorms kit. In doing so, we've used many references for inspiration. For those interested, we suggest the following references: the LEGO Mindstorms Constructopedia (versions 1.0, 1.5, and 2.0); the LEGO Extreme Creatures Constructopedia, the LEGO RoboSports Constructopedia, *Robotic Explorations* by Martin; *Building Robots with LEGO Mindstorms* by Ferrari, Ferrari, and Hempel; *Jim Sato's LEGO Mindstorms* by Sato; *LEGO Mindstorms Idea Book* by Nagata; and *Creative Projects with LEGO Mindstorms* by Erwin.

In many of his books Henry Petroski, professor of Civil Engineering at Duke University, defines *design* as be the obviation of failure. He points out that the design process is often iterative as the engineer struggles to prevent the various modes of failure. You don't learn very much from a design that works the first time around. And even if it works under one specific set of conditions (i.e. at home), that doesn't imply it will work under all conditions (i.e. in class). It's the design that falls apart that teaches you the real lessons. Knowing what went wrong is often more important than not knowing what went right. The point being, **don't expect to get it right the first time around.** LEGO bricks are great because they allow you to improve upon failed designs quickly. Take advantage of this and improve your design through several iterations.

1.4.1 General LEGO Building Tips

Before we get into the details, let's establish some of the nomenclature for LEGO bricks. Clockwise from the top, left in Figure 1.13 we have a 1x6 plate, a 1x6 throw arm, a 1x6 beam, a 1x2 axle brick, a 2x4 brick, and 2x4 plate.

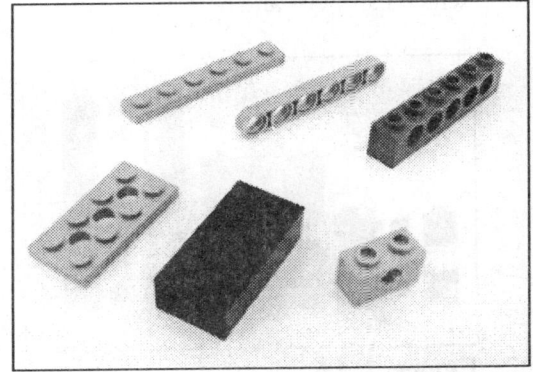

Figure 1.13
Common LEGO bricks.

Just as bricks and plates are denoted by the number of studs, LEGO axles are referred to by their length in studs. Figure 1.14 illustrates axles of length 4, 6, and 8.

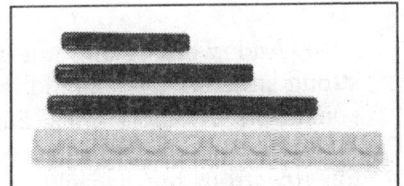

Figure 1.14
LEGO axles with lengths 4, 6 and 8.

You may have noticed that a 1x1 LEGO brick is not a cube – it is slightly taller than it is wide. In fact it is 6/5 (1.2) times taller than wide. If we designate the width of a 1x1 brick as the standard Unit, then the height is 1.2 Units.

You may also have noticed that takes 3 LEGO plates to equal the height of one brick. Thus, a plate is 0.4 Units thick. Doing a little math, it's pretty easy to see that the height of 1 brick + 2 plates equals 2.0 Units. Likewise, 3 bricks + 1 plate = 4.0 Units and 5 bricks = 6.0 Units.

Figure 1.15
A 1x1 LEGO brick is 1 Unit wide and 1.2 Units tall.

The somewhat strange LEGO dimensions turns out to be the key in understanding how to connect vertical LEGO beams to your creations. The vertical spacing between holes has to be an integral number of our LEGO Units in order for the holes to line up. Figure 1.16 illustrates assemblies with 2, 4, and 6 Units.

Why use vertical beams? The great modularity of LEGO bricks also is their largest drawback – they tend to fall apart easily. Strengthening LEGO structures with vertical beams is commonly referred to as **bracing**. To build a strong robot, which is very important in head-to-head competitions, you will need to get proficient at bracing structures.

Likewise, it is also a good idea to use at least 2 studs to join LEGO bricks. As depicted in Figure 1.17 overlapping bricks by only one stud (left) can lead to structures which can twist and deform.

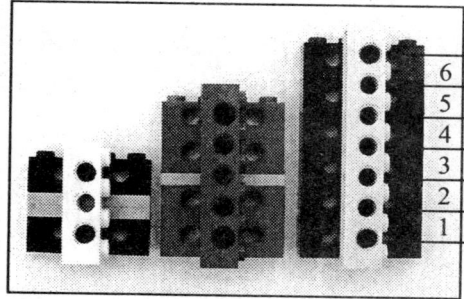

Figure 1.16
Vertical LEGO spacing: 2 Units (left), 4 Units (middle), and 6 Units (right).

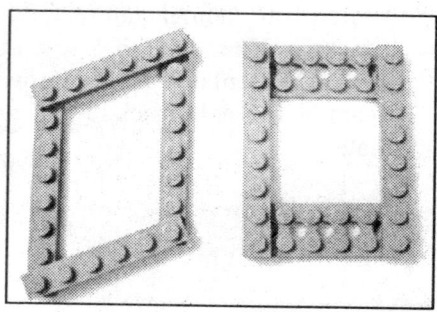

Figure 1.17
Building stiff LEGO structures.

And while we're on the subject of building strong structures we should talk about the plethora of connectors available: pins, bushes and joiners. Figure 1.18 is a photograph of the more common pins; in the top row from left to right: long friction pin, friction pin, pin, ¾ pin, ½ pin, and axle pin. Bottom row: long axle pin, long pin w/bush, bush, ½ bush, and double pin. Figure 1.19 shows the more common LEGO joiners. From left to right: axle joiner, perpendicular axle-pin joiner, perpendicular axle joiner, perpendicular axle-axle pin joiner, long perpendicular axle-pin joiner, #1, #2, #3, #5 and #6 joiners.

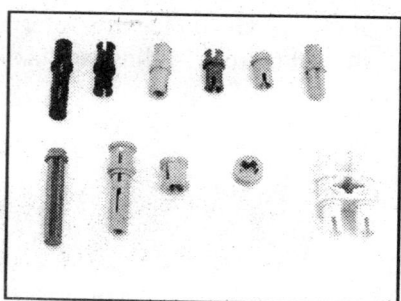

Figure 1.18
LEGO connector pins and bushes.

Figure 1.19. LEGO axle and pin joiners.

What is the difference between the black friction pins and light gray pins in Figure 1.18? The black friction pins are slightly larger in diameter than the gray pins, which makes them good for forming rigid connections. For example, in Figure 1.16 we used the black friction pins to hold the assemblies together. The gray pins being slightly smaller make good pin joints for when rotation is desirable as shown on the right in Figure 1.20. Similarly, the dark gray ¾ pins are slight smaller in diameter than the light gray ½ pins, which makes the ¾ pins good pivots for connecting beams and throw arms (1/2 width beams) as shown on the left in Figure 1.20.

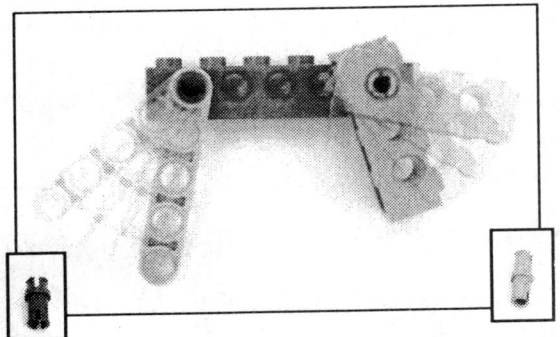

Figure 1.20
Dark gray ¾ pins and light gray pins are used as pivots between beams and throw arms (left) and beams (right) respectively.

In addition to bracing and pivots, being able to build sideways at right angles is also a handy skill. Figure 1.21 shows two methods for doing this. On the right we have used a right angle plate and on the left we have employed a few light gray ½ pins. The configuration on the left works because the stud end of the ½ pin has the same dimensions as the studs on the top of a standard brick. Neat, eh?

Figure 1.21
Two methods for building at right angles.

1.4.2 Gears and Axles

A thorough knowledge of gears and axles is also crucial to building a LEGO robot. At full power the standard LEGO motor (Table 1.1) spins at approximately 350 rev/min, which may be fine for some designs but is probably too fast for most applications. Gears are used to change the rotational speed (rev/min) and torque between shafts or axles.

Figure 1.22. Common LEGO Gears (left to right):
Top row: 40T spur, 24T spur, 16T spur, 8T spur, differential
Middle row: 24T crown, 24T double bevel, 12T double bevel, 12T bevel
Bottom row: 24T clutch, linear rack, worm, 16T idler, 4T axle, 8T knob.

Strictly speaking an axle is not a gear, but we have used them as 4 tooth gears in a pinch. Likewise the 8T knob is not meant to be a gear but can be used if necessary.

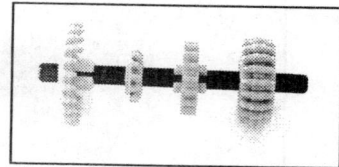

Figure 1.23. Types of gears.
Left to right: crown, bevel, spur and double bevel. Only the 12T bevel gear is ½ a Unit thick; all the other gears are 1 Unit thick.

Figure 1.24. Turning corners.
Both crown and bevel gears are used to transfer power between two axles that are perpendicular to each other.

If we ignore friction (which isn't always such a good assumption, but we'll proceed anyway), we can express the relationship between the product of **number of teeth**, n, and **angular velocity**, ω, at one gear to the number of teeth and angular velocity of a second gear as:

$$n_1 \omega_1 = n_2 \omega_2 \qquad \text{(eq. 1.1)}$$

If the big gear is driving the little gear, this is called "gearing up" because the small gear will spin faster than the big gear. Conversely, when a little gear is driving a big gear, this called "gearing down." As shown in figure 1.25, the 40T gear will spin 0.6 times as fast as the 24T gear.

There is also an equation that relates the product of the **torque**, T, and **angular velocity**, ω, at one gear to the torque and angular velocity of a second gear. If you are not familiar with these terms, you can think of torque as the twisting force on a gear or axle and the angular velocity as the rotational speed (e.g. revolutions per minute). The torque and angular velocity of two mating gears are related by the following equation:

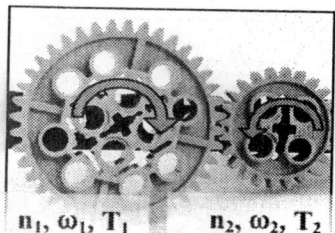

Figure 1.25
Gear relationships

$$T_1 \omega_1 = T_2 \omega_2 \qquad \text{(eq. 1.2)}$$

Combining Equations 1.1 and 1.2 leads to:

$$T_1 n_2 = T_2 n_1 \qquad \text{(eq. 1.3)}$$

Since $n_1 = 40$ and $n_2 = 24$, the torque on the little gear's axle will be 0.6 times the torque on the big gear's axle in Figure 1.25.

Figure 1.26. Gear ratios.
Both arrangements use 8T and 24T gears.
LEFT: low speed, high torque. The output shaft will turn 1/3 as fast as the motor. Because the arrangement can produce a lot of torque, we've used the white clutch gear to prevent stalling the motor.
RIGHT: high speed, low torque. The output shaft spins 3 times as fast as the motor, but will produce very little torque.

Figure 1.27. Gear radii and gear trains.
The 40T, 24T, and 8T gears are 2.5, 1.5, and 0.5 Units in radius respectively. When used together, their centers are an integer number of Units apart and can be easily meshed on a single LEGO beam. Conversely the 16T gear (not shown) has a radius of 1.0 Units and, thus, will only mate to another 16T gear when used on a single LEGO beam.

Using Eqn. 1.1 the 8T gear will spin 3 times as fast as the 24T and 5 times as fast as the 40T gear. Note the 24T gear will spin in the opposite direction as the others. The 8T gear will require 1/5 the amount of torque to turn as the 40T gear according to Eqn 1.3.

Figure 1.28. Compound gear trains.
On axle #2 we have used both a 24T and an 8T gear. In doing so we have created two joined gear trains, one between the axles 1 and 2 and the second between axles 2 and 3. This is called a **compound gear train**. Using Eqn 1.1 we find that $\omega_1 = (^{24}/_8)\omega_2$, thus, the 8T gear on axle #1 will spin 3 times faster then the 24T on axle #2. Both the 8T and the 24T gears on axle #2 will spin at the same angular velocity since they share the same axle. Again using Eqn 1.1, $\omega_2 = (^{40}/_8)\omega_3$. Combining these results we find that the 8T gear on axle #1 spins 15 times faster than the 40T gear: $\omega_1 = 15\omega_3$.

Using Eqn 1.3 twice, we also find that $T_1 = (^1/_{15})T_3$.

Figure 1.29. Too much torque.
Here's a good argument for using the white clutch gear. Through the use of compound gear trains is not very difficult to get an axle which is spinning extremely slowly and, thus, capable of producing a lot of torque. In this figure the LEGO axle was obviously subjected to more torque than it could handle.

Figure 1.30. Worm gears.

Worm gears are a special type of gear in which the axles are perpendicular to one another. The worm gear is a 1 tooth gear and, thus, is commonly used to greatly reduce the angular velocity and greatly increase torque. Another feature about a worm gear is that the worm gear can drive the pinion (the spur gear), but the pinion cannot drive the worm gear - it binds up.

LEGO makes a neat little gearbox (right), but it's also fairly easy to build one yourself if you want to use a worm gear. As shown the worm gear will spin 24 times slower than the spur gear.

Figure 1.31. Aligning multiple worm gears.

If you want to stack several worm gears in a row, make sure you have the orientation correct or it won't work. In the figure we have shown a correctly (top) and incorrectly (bottom) aligned pair of worm gears.

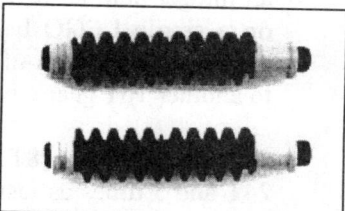

Figure 1.32. Linear motion parallel to axle.

The worm gear can be used to convert rotational motion into linear motion. Each complete revolution of the axle will advance the rack one tooth. Notice the rack will move parallel to the axle when a worm gear is used.

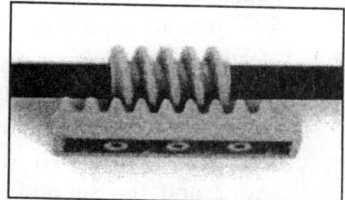

Figure 1.33. Linear motion perpendicular to axle.

Linear motion can also be accomplished with a rack and pinion set up. Unlike the worm gear, the motion is both perpendicular to the axle and much faster.

Figure 1.34. The differential.

The differential is unique in that it is a gear train with 3 input/output axles. In your car axle 1 would be driven by the motor and the wheels would be connected to axles 2 and 3. The differential allows the two wheels to spin at different velocities as your car goes around a corner (the outside wheel must spin faster since it travels a farther distance). The book, *The Way Things Work*, and the accompanying website (www.waythingswork.com) provide an excellent description of a differential.

Figure 1.35. Pulleys and belts.
Pulleys act very much like gears with the exception that slipping may occur between the belt and the pulley. One of the biggest advantages of using pulleys is that the various sizes and elasticity of the belts affords a lot of flexibility in locating the axles. You are not restricted to spacing pulleys an integral number of Units apart, as is the case with gears. Eqn 1.1 works if you replace the number of teeth with diameter. Eqn 1.2 holds true up until the belt starts to slip on one of the pulleys.

NOTE: Belts are NOT rubber bands. They are not intended to be stretched very much and will break easily if stretched even moderate amounts.

> Stalling a motor will drain the batteries FAST. To prevent stalling a motor you may want to use pulleys and belts instead of gears. The 24T clutch can also be used (figure 1.22)

1.4.3 Getting Around on Wheels

Let's now turn our attention to the art of moving around. The two most basic methods of building robots that move are to use either wheels or legs. Creating robots with wheels is much easier for most people, so we'll start there first.

Figure 1.36. Common LEGO wheels and tank tread.
LEGO provides a literal plethora of wheels to work with. Shown are some of the wheels that are commonly found in Mindstorms kits.

Figure 1.37. Tank treads.
Tank treads are commonly used for locomotion. The typical setup involves using a 16T spur gear inside the white tread hub. The upper photo shows an 8T spur gear mounted on the motor output shaft driving a 24T spur gear. The 16T spur gear is mounted on the same shaft as the 24T gear.

Vehicles that use two tank treads can turn by spinning the treads in the opposite direction. This is commonly called "skid steering."

Figure 1.38. Triangle tank treads.
Here's an alternative arrangement that uses three tread hubs. It also makes a nice conveyor belt.

Figure 1.39. Four wheel drive.
The biggest drawback of the skid steering is that it has a lot of friction and, thus, tends to be very slow. However, the tank treads can be replaced with closely spaced wheels, which results in a very maneuverable and fast method of locomotion. It's still skid steering, but it's better than tank treads

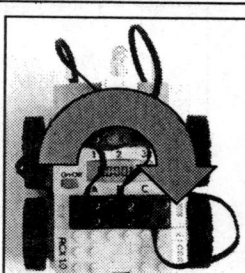

Figure 1.40. Skid steering.
Turning the motors in opposite directions results in a robot that turns on the spot (zero turning radius).

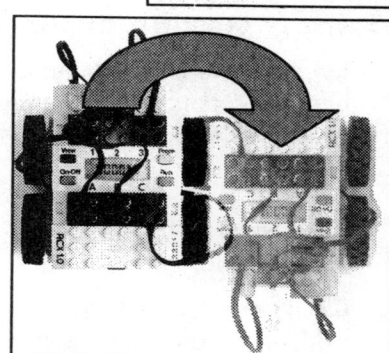

Figure 1.41. Turning on the left motor while stopping the right will result in right turn that pivots about the right wheels.

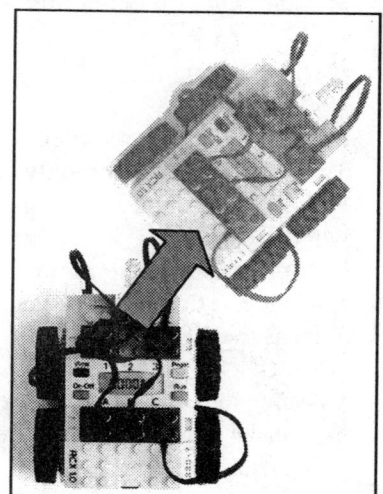

Figure 1.42. Running the left motor faster than the right results in a large turning radius right turn.

  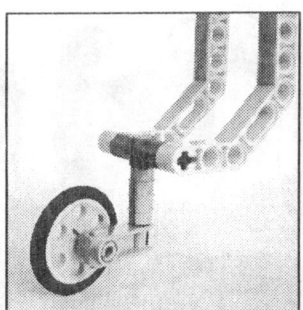 

Figure 1.43. Caster wheels.
Caster wheels are the type of swiveling wheels on the front of a shopping cart. Caster wheels are useful for making two wheel drive robots. Several typical arrangements are shown.

Figure 1.44. Friction button.
Even better yet, the caster wheel can be completely replaced with a friction button and only two wheels are needed for the car.

Figure 1.45. Sample two wheel cassis.
This figure shows top and bottom views of a simple two wheel cassis that uses two friction buttons for balance. Because the wheels are connected directly to the motors, this vehicle is very quick.

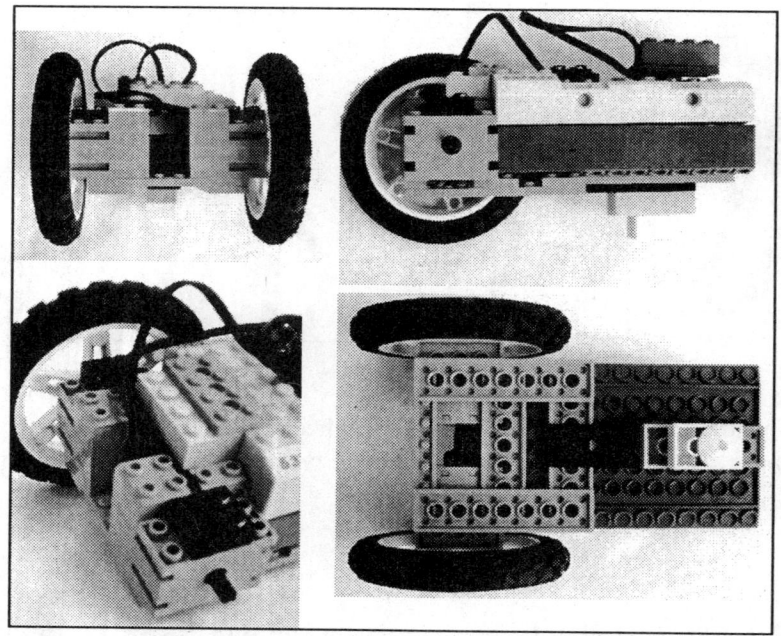

Figure 1.46. Another two wheel chassis.
This figure shows three orthogonal views and detail view of the motor mounts for another simple two wheel cassis. Again a friction button (white) is used for balance.

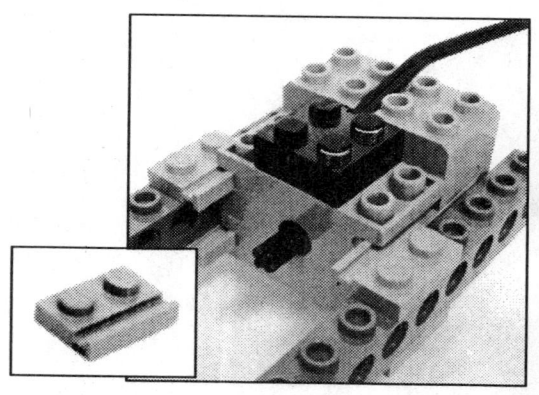

Figure 1.47. Mounting motors.
The LEGO motors are surprisingly powerful. Mounting a motor securely can be accomplished with the 1x2 motor mount plates as shown in the upper two figures. An alternative motor mount, shown on the right, utilizes 2x6 plates both above and below the motor.

Figure 1.48. One motor turning.
If you use two motors for driving, that doesn't leave many options for claws. One solution to this dilemma is the use of the differential. As shown, one motor is used for both driving and turning. When the car goes forward (to the right) it drives straight. When it backs up, the ratchet locks the bottom right wheel and the car turns.

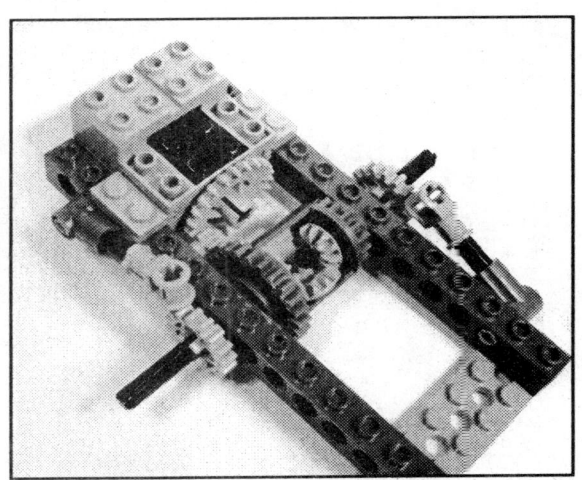

Figure 1.49. Output selector.
In this photo, we've used the differential and 2 ratchets, like the one shown in Figure 1.48, to ensure both axles never spin at the same time. When the motor spins forward, only one of the output axles turns. Reversing the motor direction causes the other output axle to turn. Two outputs with one motor! We've seen plans for a 5 output selector, but it's a bit beyond the scope of this book.

Figure 1.50. Worm gear output selector.
In this photo, we've used a worm gear to select one of two output axles. When the worm gear axle (the input) is turned one way the right output shaft spins (top). Turn the input axle the other way, and the worm gear moves left and causes only the left output shaft to spin (bottom). The bottom two gears are only used to supply a little bit of friction that is needed for this to work.

Figure 1.51. Belt Drive.
Unlike gear trains, pulley and belts allow for a variety of arrangements because the belts are flexible (don't stretch them too far). Pulley and belt drives also don't require very rigid motor mounting.

1.4.4 Getting Around on Legs

Vehicles with wheels are more prevalent than vehicles that utilize legs, mainly because legs tend to be more difficult to implement. However, we have found that it's the robots with legs that people find most intriguing. To help you develop your leg-building skills, this section will cover some of the more common techniques used.

An important concept to remember is that when a leg lifts off the ground, the robot must either balance on the other leg(s) or have some kind of support. In figure 1.52, we have employed stationary supports behind the legs. In figure 1.53, no supports are shown. If the robot does not have any supports, then it will roll on the gears resulting in an awkward combination of walking and rolling.

Figure 1.52. Basic legs.
The legs will appear to move in an elliptical path. Supports must be used to keeps the robot balanced when the feet lift off the ground.

Figure 1.53. Variation on the basic legs.
Here we've introduced a cosmetic change to the basic legs. These legs behave the same as the legs in the previous figure, but look very different. No supports are shown, but must be used for balance.

Figure 1.54. Four-bar linkage legs.
Slightly more complicated than the basic legs, four-bar linkage legs can be used to create a wide variety of leg motion. Here we use a parallelogram-type design. The foot will move in a circular path. The gears on the back keep the links synchronized.

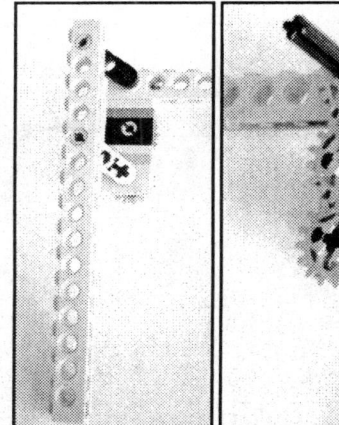

Figure 1.55. Unequal link lengths.
This four-bar linkage design utilizes 4 links of different lengths (the 40T gear serves as one of the links). The resulting motion is elliptical.

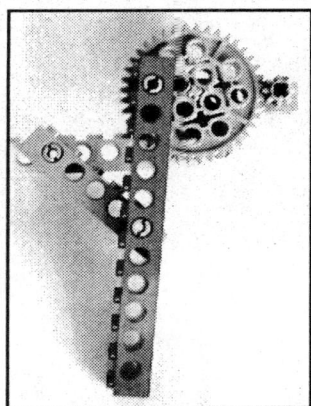

Figure 1.56. Reciprocating legs.
This leg design is similar to the piston-cam mechanism in most cars. The "foot" follows an elliptical path. The length of the leg, the diameter of the gear, and the distance to the brace all affect the foot path. At least three legs are needed for each side of a robot for balance. The Internet is filled with various hexapod (6-legged) robot designs based on this basic format.

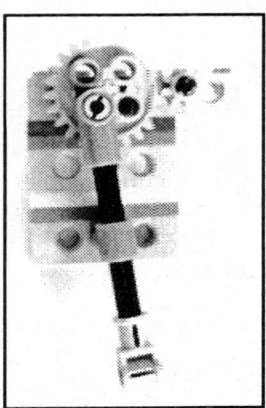

Figure 1.57. Hexapod.
To build a hexapod, timing of the legs is critical. Each step, the robot must lift up 3 legs at a time. In this figure, only the center leg is lifted. On the opposite side of the robot, only the center leg would be down (2 legs off the ground).

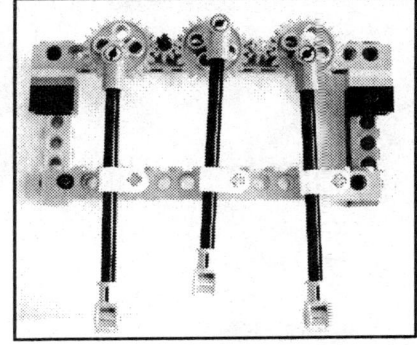

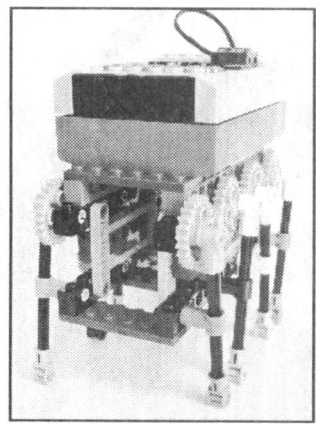 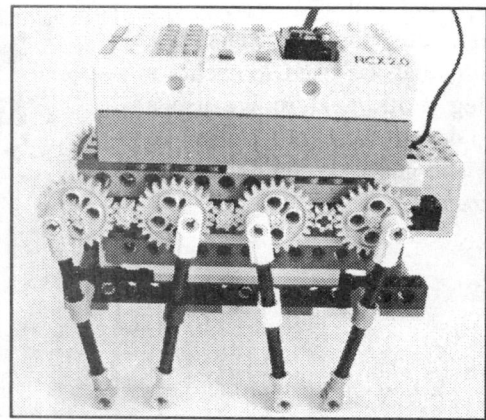 

Figure 1.58. Putting it together.

As an example of putting it all together, we've constructed a simple eight-legged robot. Notice the vertical bracing on the inside of the chassis (far left figure). This robot lifts 4 legs at the same time (2 on each side).

Figure 1.59. Best legs contest.

Here are some ideas for decorating your robot's legs

Figure 1.60. Balancing robot.

We've left off the RCX, but this simple single motor robot can balance on one leg as it lifts the other. The result is very realistic gait. A worm gear has been used to gear down the motor output.

1.4.5 Bumpers and Sensors

Being able to make a robot that reacts to its environment is one of things that makes the Mindstorms product line so interesting. In this section, we present some ideas for bumpers using the touch and light sensors.

Figure 1.61. Simple bumper.
The figure shows a simple bumper that uses a single axle to active the touch sensor.

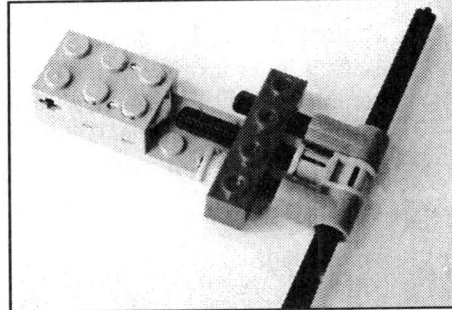

Figure 1.62. Guided bumper.
This simple bumper uses two axles as guide pins to prevent twisting and turning. Note the ½ bush used on the lower guide axle to retain the bumper.

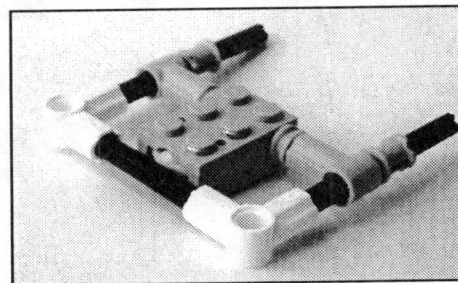

Figure 1.63. Guided bumper #2.
Here's another simple bumper that uses an axle passing through the axle hole in the touch sensor and two #1 pin joiners as guides.

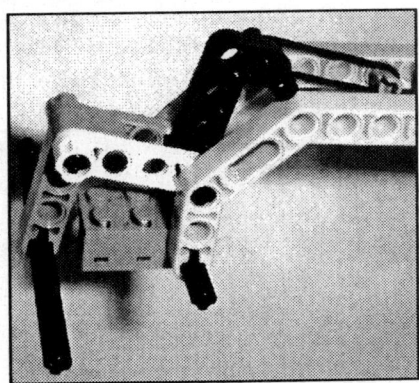 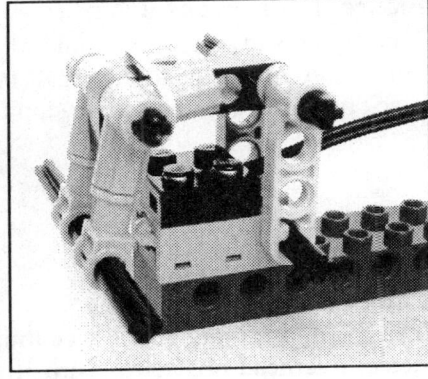

Figure 1.64. Lever bumpers.
The figures show two examples of lever-arm type bumpers. Both bumpers utilize rubber bands to hold them open rather than relying on the internal spring inside the touch sensor.

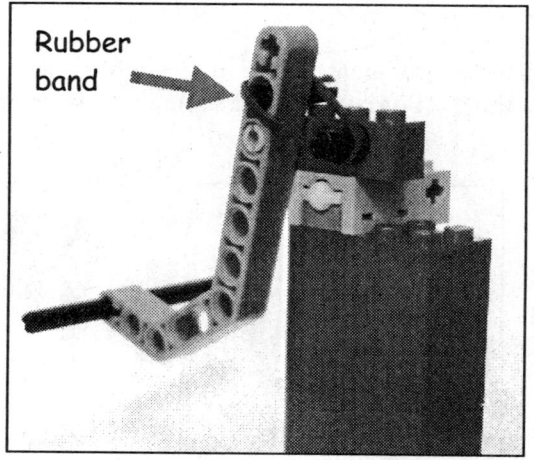 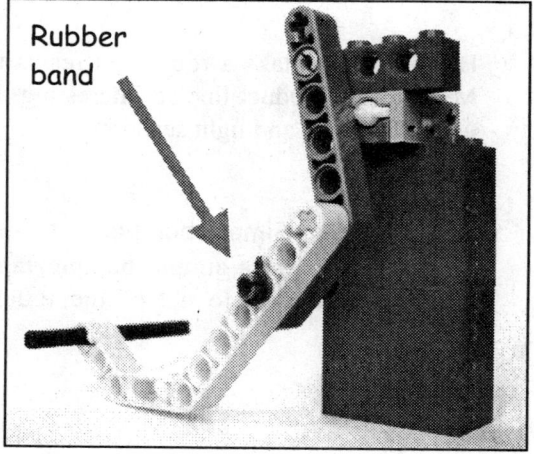

Figure 1.65. Lever arm bumpers.
Left: a lever arm activates the touch sensor when it swings back. A rubber band is used to return the arm to the neutral position shown.
Right: a bumper that reacts to both bumps and dips. A rubber band between the two arms provides the return force.

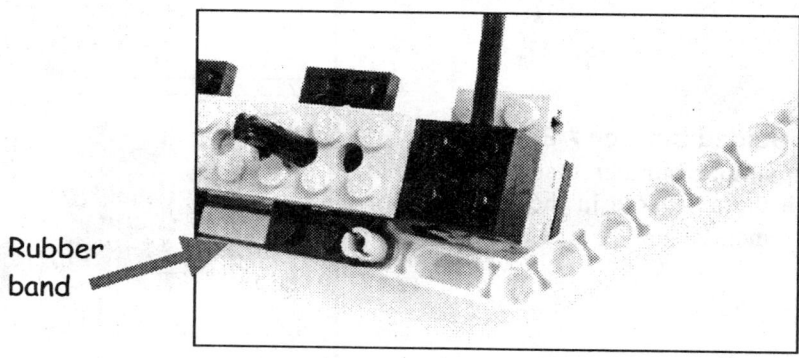

Figure 1.66. Front Bumper.
This figure shows the right side of a lever arm bumper that can be used on the front of a vehicle. A rubber band stretched between the two lever arms keeps them open. The 2x4 black plate serves as a positive stop for the lever arm.

Figure 1.67. Pivot bumper.
This bumper pivots about a central axle. Use as shown to detect overhead obstacles. Turn it upside down to use as a traditional bumper.

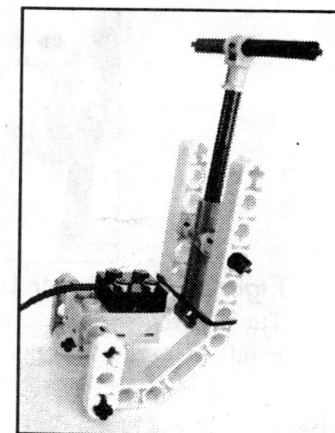

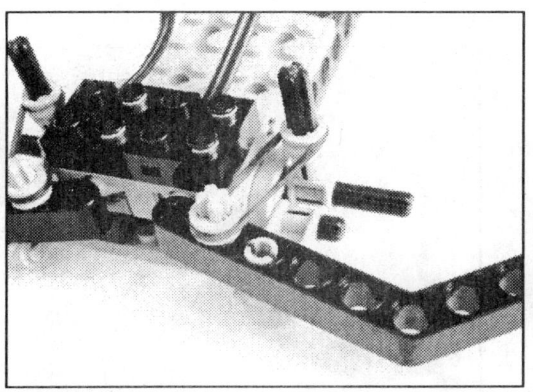  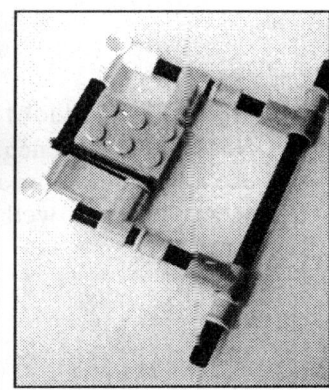

Figure 1.68. Normally closed bumpers.
Most bumpers rely on the touch sensor being pressed to activate. These three bumper designs rely on the touch sensor being released to activate. All three bumpers rely on a small rubber band to keep the touch sensor normally closed.

Figure 1.69.
Complex normally closed bumper.
Here's more complex normally closed bumper. It relies on a rubber band across the diagonal of a four bar linkage to keep the touch sensor depressed. The bottom figure shows the back side with one of the throw arms removed while being activated (opened).

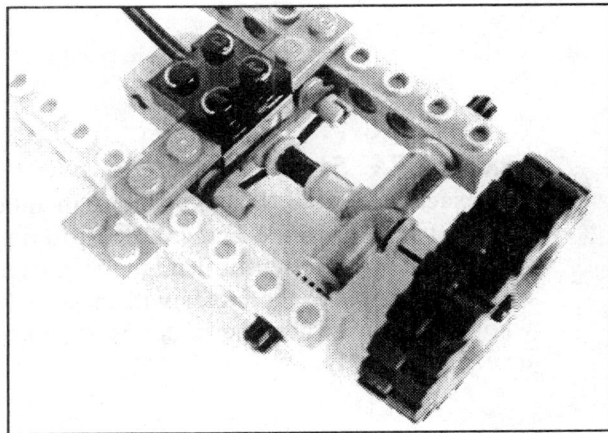

Figure 1.70. A wheel bumper.
A slight modification to the earlier pivot bumper, this design is a normally closed bumper that reacts to the up and down motion of a wheel. The rubber band keeps it centered. This design can also be used a simple joystick control.

Figure 1.71. Light sensor bumper.
Need an extra touch sensor? Here's how to use a light sensor as a touch sensor. This only seems to work well for dark colored lever arms.

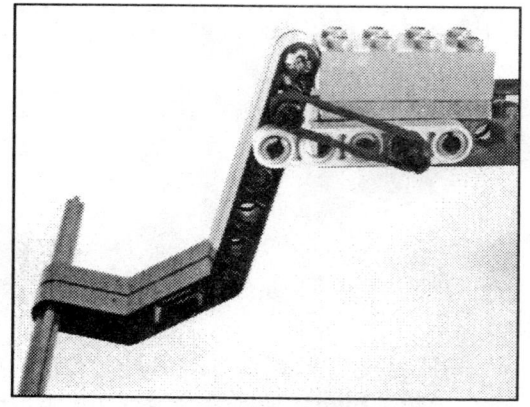

Figure 1.72. Light rotation sensor.
Here we've used the light sensor to measure rotation. With six positions, this sensor has a resolution of 60 degrees.

Figure 1.73. Touch rotation sensor.
The touch sensor can also be used to sense rotation. Here a cam is used to activate the touch sensor once per revolution.

Figure 1.74. Stepper motor.
Not exactly a sensor, but this simple method can be used to create accurate positioning without the use of a sensor. Due to the tension in the rubber band, the motor will make only whole rotations when the motor is switched on and off quickly.

1.4.6 Grippers and Claws

Now that you know how to move around and sense that you've bumped into something, we will discuss how to grab onto things. There are two main types of claws/grippers: motor actuated and over-the-center actuated. Motor activated claws use, not surprisingly, a motor to provide the gripping force. Over-the-center grippers use a rubber band that is triggered by some kind of switch.

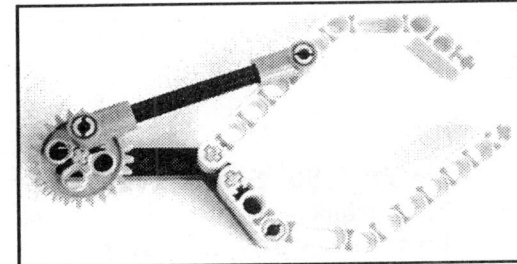

Figure 1.75. Basic gripper.
The typical gripper uses a gear or cam, powered by a motor, to open and close the claws.

Figure 1.76. Cam gripper.
This gripper looks slightly more complex, but mechanically it is the same as the basic gripper. The lower claw is held stationary with a rubber band.

Figure 1.77. Scissor claw.
This claw uses the classic scissor mechanism actuated by a 40T gear. Changing the lever arm lengths will affect the range of motion of the jaws.

Figure 1.78.
Modified scissor gripper.
This gripper is mechanically equivalent to the scissor claw, but uses lever arms instead of beams. We also used the 24T white clutch gear so that the motor isn't stalled when we grab things (stalling a motor drains the batteries very fast).

Figure 1.79. Worm-gear claw.
This claw is actuated by a worm gear. The two spur gears also serve to keep the upper and lower jaws synchronized. Because of the high gear reduction, this jaw can produce a lot of gripping force to destroy your opponents!

Figure 1.80. Synchronized gripper.
This gripper has two unique features. First, it uses 3 gears to synchronize the motion of the upper and lower jaws (far right). Second, it uses a rubber band to hold the gripper either open (center) or closed (left).

Figure 1.81. Over-the-center claw.
With an over-the-center claw, the jaws are held open by a rubber band. A throw arm acting against a pin keeps the jaws from opening too far. The claw closes when the trigger is pressed. This causes the upper jaw to rotate slightly and then the rubber band snaps the jaws closed.

Figure 1.82. Another OTC claw.
This over-the-center claw is a much cleaner design and utilizes a linear-motion trigger. Again, the rubber band changes from holding the jaw open to snapping it closed when the trigger is pressed.

1.4.7 Creativity & Aesthetics

Creativity and aesthetics are often overlooked in engineering, but both play a crucial role in the design process. The truth of the matter is that how well a product does in the market is often more governed by its form than its function. While we don't support designing for looks alone, it certainly cannot hurt to have it mind during the design process. Furthermore, creative designs typically cost less and do better in the marketplace than non-creative ones.

And, contrary to popular belief, both creativity and aesthetics are not innate skills reserved for only the "artistic." There are several good references that discuss structured creative thinking processes such as **brainstorming**.

The Art of Innovation by Tom Kelly is very good at providing some insight as to the importance of creativity and aesthetics. The book discusses the way IDEO, probably the most famous "innovative" design firm in the world, goes about its business. IDEO is responsible, among other things, for the original Apple computer mouse, the Polaroid i-Zone instant camera, and the Palm V personal digital assistant. IDEO is a very successful company (by almost any measure of success) because it fosters and rewards creative thinking and demands attention to aesthetics from its employees.

Creative Blocks

So how do you go about thinking creatively, even if you don't consider yourself creative? **Structured creative thinking.** *Conceptual Blockbusting* by James Adams and *A Whack on the Side of the Head* by Roger von Oech are both excellent books that cover techniques for creative thinking. Both books spend a great deal of time emphasizing ways to overcome "creative blocks." Adams defines creative blocks as:

> *"mental walls that block the problem-solver from correctly perceiving a problem or conceiving its solution."*

Creative blocks are necessary for surviving everyday life. The concept of driving on the left hand side of the road in the U.S. may be creative, but not very safe. Seeing someone walking out of a hospital wearing scrubs usually leads to the conclusion that he works at the hospital. We reach this type of decision without all the facts everyday. The truth, however, may be that he is a mental patient trying to escape!

Anyone who has children quickly realizes that children are both extremely creative and extremely accident prone. I think the two go hand in hand. When was the last time you tasted something to determine its function?

Oechs lists ten common creative blocks:
1. Looking for the one "right answer"
2. Being too logical
3. Following the rules
4. Being too practical
5. Believing play is frivolous
6. Sticking to your own area of expertise
7. Being afraid to look foolish
8. Avoiding ambiguity
9. Believing "to err is wrong"
10. Believing you're not creative

Looking for the "right answer" is a great way to stifle creativity because you presume there is only one solution. However, the most obvious solution is usually not the best.

Being too logical tends to elicit comments like "that won't work because..." or "that doesn't make sense because..." In order think creatively, you often have to examine what initially seems outrageous or implausible.

Following the rules is usually a good habit, however, most rules have a gray area and it is these areas which can be exploited to achieve creative solutions. In my class, I encourage as much "cheating" as possible in the design competitions. By interpreting the rules as loosely as possible, you'll have more solutions available to you. However, be forewarned that you always run the risk of being disqualified for breaking the rules!

Work should be fun. A great brainstorming session is always punctuated by lots of laughter. Have fun at the start of the design process. Think of lots of crazy ideas. Then use logic and math (your engineering skills) to reduce those ideas to a workable solution.

Recruiting the opinion of someone from the outside is a great way of spurring creative solutions because they don't know what the "correct" solution is or "how it's always done." Likewise, when you are working outside your area of expertise, you don't know what is and isn't possible. That's one reason why kids can be so creative – they don't know what's impossible!

I mentioned earlier that laughter is a sure sign of a good brainstorming session. It goes without saying this means you have to be comfortable having your ideas laughed at once in a while. It's often the laughable concept that ends up being the best in the end.

Engineering and science disciplines stress eliminating ambiguity – precision is important. However, during the initial design phases, ambiguity is cherished because it doesn't constrain a solution. "Lets use a big shaft" doesn't specify the size of the shaft. Only once the detail design phase starts do we want to start getting that specific.

"You learn from your mistakes." It's okay to make mistakes once in a while. Brainstorming sessions are not about pointing out technical infeasibilities – they are about generating creative solutions.

If you don't truly believe you're creative, you won't be. Everyone can be creative with a little practice.

Structured Creativity

Brainstorming is the most popular form of structured creativity. The goal of the brainstorming session is to generate as many solutions to the problem as possible. A typical brainstorming session only lasts about 5 minutes. The rules for a brainstorming session typically look like the following:

- No criticizing. Don't judge anyone's ideas during the brainstorming session.
- No hesitating. No idea is to be considered too outrageous, ridiculous, or "wacky."
- No holding back because you have "too many" ideas. The more the better.
- No owning of ideas. If you can build upon or improve upon someone else's idea, write it down.
- No sticking to your own field of expertise. Think globally across disciplines.

Brainstorming may be the most common structured creative thinking technique, but it is far from the only one. **List making** in one form or another is also popular technique. For example, if you are trying to find new uses for a brick, listing all its attributes (red, brittle, rough, heavy, etc) and then determining ways to exploit each individual attribute will lead you to lots of "creative" uses. One year a student suggested starting a new Internet company called www.protestersupplies.com.

By using some form of structured creative thinking, you can overcome creative blocks. When this happens, a whole new classes of solutions is possible. I witnessed a great example of creative thinking (not on my part unfortunately) at a workshop I attended for new faculty. During one part of the workshop we were asked to make a device using LEGO bricks. I won't get into the specific objectives of the task, but suffice to say it required some creative solutions given the limited assortment of LEGO bricks we were provided with. Someone on another team got the brilliant idea to cut the teeth off an 8T spur gear to make a watch-like mechanism. The photo shows a gear train with a 40:1 gear ratio that has intermittent motion. The concept of destructively modifying a LEGO gear never occurred to me. However, once the other team overcame this creative block, a whole new class of solutions we available to them. As you can imagine, this provided them with a distinct competitive advantage because they had a function that nobody else had.

Figure 1.83 Creative modification of an 8T LEGO gear.

1.5 Online Resources

There are also a lot of very good websites that you can also visit to get more information. Here are just a few sorted by topic.

Official ROBOLAB Materials and Information

- Official Mindstorms for Schools Website - **http://www.lego.com/eng/education/mindstorms**. This web site features information on ROBOLAB products as well as links to official downloads and support resources.

- Tufts University, the maker's of ROBOLAB - **http://www.ceeo.tufts.edu/Robolab**

- Download software patches an other resources at - **http://www.ceeo.tufts.edu/robolabatceeo**

- National Instruments, the maker's of LabVIEW **http://www.ni.com/company/robolab.htm**

- The official Mindstorms LEGO site- **www.lego.com/eng/education/mindstorms** This site is for the retail version of Mindstorms, which is different than the Mindstorms for Schools sets used in this book.

Where to buy

- Pitsco LEGO - DACTA Store - **http://www.pldstore.com**
  The Pitsco LEGO-DACTA store is the official retailer of all LEGO Educational Materials in the U.S. ROBOLAB can be purchased from here by any consumer (teacher, parent, individual etc.)

- LEGO World Shop – **http://shop.lego.com**
  The official LEGO online store.

- Mondo-tronics, inc – **http://www.robotstore.com**
  Roboto Books.com – **http://www.robotbooks.com**
  Two great source of all things robotic

- Unofficial Online LEGO shops where you can buy extra bricks
  - **http://www.bricklink.com**
  - **http://www.brickshelf.com**

Groups

- LUGNET (Lego Users Group Network) - **http://www.lugnet.com**
  LUGNET has fabulous LEGO resources ranging from pieces in sets to a wide array of discussion groups on robotics, education, ROBOLAB and more.
     Discussion group on LEGO education - **http://news.lugnet.com/edu/**
     Discussion group on LEGO dacta - **http://news.lugnet.com/dacta/**
     Discussion group on LEGO robotics - **http://news.lugnet.com/robotics/**

- MIT's Epistemology and Learning Lab - **http://el.www.media.mit.edu/groups/el/**
  Check out where the whole programmable brick idea came from and see where it might go next.

Resources

- NQC (Not Quite C) - **http://www.enteract.com/~dbaum/nqc**
  If you want to program the RCX with a C-like syntax than NQC is an excellent alternative to ROBOLAB.

- LEGO SDK **http://mindstorms.lego.com/sdk/SDK.asp**

- Java **http://lejos.sourceforge.net/**
  New Java interface for programming the RCX

- LegOS, **http://legos.sourceforge.net/**

- NASA's Robotics Education site - **http://robotics.arc.nasa.gov/**

- NASA's LEGO Data Acquisition and Prototyping System - **http://ldaps.arc.nasa.gov/**
  See the precursor to ROBOLAB and lots of interesting projects.

Examples, Ideas and Cool Projects

- Boulette's Robotics in Luxemburg (**http://www.convict.lu/Jeunes/RoboticsIntro.htm**) A fabulous site that features high end robotics and ROBOLAB projects with great descriptions and the code used.

- Michael Gasperi's excellent page on "extreme mindstorms," including great tip on building sensors - **http://www.plazaearth.com/usr/gasperi/lego.htm**

- Doug's LEGO robotics page. Lots of neat projects - **http://www.visi.com/~dc/**

1.6 LEGO Design Challenges

It is often said that we remember only 10% of what we are told but 90% of what we teach others. There is also an old Chinese proverb that goes something like this: "Tell me and I forget. Show me and I remember. Involve me and I understand." Combined, these capture our teaching philosophy and the approach we've taken in this book.

In developing the design challenges for this book, we've assumed you are working together on small teams. By working on small teams, you discuss, argue, and fight about the solutions. This interaction gets you directly involved in the learning process and makes you think about your approach. The instructor becomes more of a facilitator than a teacher.

As for putting the challenges at the beginning of the chapter (with the exception of this first chapter), we realize most students will start with by reading the problem and then commence scanning the chapter for the equations they need. We've cut to the chase by putting the challenges at the beginning of the chapter and then listed the skills required for each challenge. We've organized the book by skills so you can skip right to the section you need to complete the challenge.

The skills badge approach also has one unique advantage over the traditional problem set. In a "normal" homework set, perhaps 95% of the comments you get back are negative in nature, highlighting your errors. The design challenges are meant to let you show off the skills you have learned, which we think is a much more constructive method of teaching. Plus, its way more fun!

1.6.1 Team Communication

Challenge: Develop oral communication skills by building a LEGO sculpture aided only by verbal instructions.

Skill Badges: None

Procedures:

Experimental Setup: Each pair of students should receive an identical set of LEGO bricks.

Robot Design:
- This is an exercise for two people. Check to make sure your Lego pieces match identically (size, shape, and color).
- Select one of the people to be the leader for this exercise.
- The two participants should sit back to back.
- The leader should design and construct a Lego object (anything he/she wants). As the leader is building the object he/she is to explain to their partner how to assemble the same object using verbal instructions only
- The second partner is not to speak or gesture in any way if there is mis-understanding – no peeking! Example: if the second partner missed the color description s/he is not allowed to say "what color?", or gesture with an elbow to repeat the comment.

The second partner cannot talk or ask questions by any means.

- Once the objects are completed the partners are to face one another and discuss the differences of appearance (if any).

Program: None

Analysis: Answer the following questions:

- How did your team decide who would be the leader?

- Did you feel confident and comfortable during the exercise?

- What were the main reasons for error?

- Did you get better as you went along?

- What might you do differently next time?

- How might you use what you've learned from this exercise in the future?

1.6.2 Drag Race

Challenge: Design and build a vehicle powered only by a rubber band.

Skill Badges: None

Procedures:

Experimental Setup: All that is needed for this project a stopwatch for timing.

Robot Design:
- This is a three-phase, timed in-class design competition.
- Your score will be based upon a combination of time to market and distance traveled.

Phase I:
- You have 30 minutes to design, build and test a rubber-band powered vehicle. The official rubber-bands will be supplied by the instructor.
- The time to market is defined as the time you 'release' your design to the rest of the class. Your vehicle must be on public display at front of class until the competition starts. No modifications are allowed once a design is 'released'. Copying of released designs is encouraged!
- The winner is determined as the longest distance traveled. The distance traveled is defined as distance from the start line to the LEGO piece closest to the starting line.

Phase II:
- You have 5 minutes to re-design your vehicle. Pay attention to designs that did well!
- The winner is determined as the longest distance traveled.

Phase III:
- You have 5 minutes to re-design you vehicle once more.
The winner is determined as the longest distance traveled.

Program: None.

Grading:

Your grade will be based 40% on time to market, 40% on distance traveled and 20% on creativity and aesthetics.

Time to market	Distance traveled	Creativity & Aesthetics
A+: First to market.	A+: Moved the farthest	A+: Best of show
A: Within 3 minutes	A: Moves 15+ feet	A: Outstanding
B: Within 5 minutes	B: Moves 10+ feet.	B: Good
C: Within 10 minutes	C: Moves forwards.	C: Okay
D: Within 15 minutes	D: Moves backwards.	D: Nothing special
F: Didn't finish	F: Didn't move	F: Divert your eyes!

1.6.3 South Pointing Chariot

Challenge: Design and build a south pointing chariot. Your cart must navigate a figure-eight course, all the while pointing south.

Skill Badges: None

Procedures:

Experimental Setup: A figure-eight course and a compass is required for this project.

Robot Design: The photo shows a South Pointing Chariot, which was invented in China in the third century AD and is the first known use of a differential gear.

A figure was mounted on the two-wheel carriage which always pointed south, no matter how the carriage turned as it moved. The figure was set to south and then a differential driven by the wheels turned the figure in the opposite direction to the carriage so that it still pointed south.

Unlike most gear trains, which have only two axles (an input and an output), a differential has 3 axles. In most applications that use a differential, like your car, a motor (the input) turns axle #1 and the wheels (the outputs) are connected to axles #2 and #3. For the south pointing chariot, the wheels will be connected to axles #1 and #3 and the pointer will be connected to axle #2. Try building up just the differential and playing with it to see how it works. Try holding one axle, while turning another.

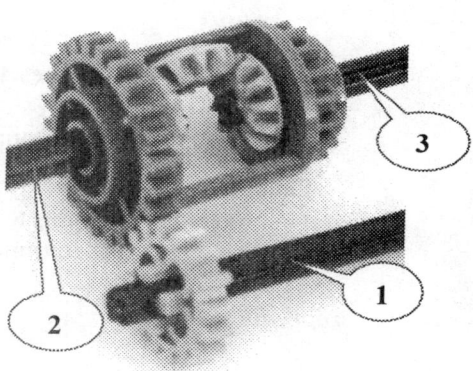

Program: None

Grading:

Your grade will be based 80% on accuracy and 20% on creativity and aesthetics.

Accuracy	Creativity & Aesthetics
A: Always points south	A+: Best of show
B: Turns the right direction, but doesn't track due south	A: Outstanding
	B: Good
C: Pointer turns	C: Okay
D: You build something with wheels	D: Nothing special
F: You don't show up to class	F: Divert your eyes!

1.6.4 Heavy Lifting

Challenge: Design and build a LEGO crane.

Skill Badges: None

Procedures:

Experimental Setup: The only specialized components required are various calibrated weights. Your instructor will indicate how the crane is anchored (if at all) to the table.

Robot Design: Gear trains and vertical bracing will be important construction skills to learn.

Program: None

Grading:

Your grade will be based 80% on the weight lifted and 20% on creativity and aesthetics.

Weight lifted	Creativity & Aesthetics
A+: Most lifted	A+: Best of show
A: More than 3 lbs	A: Outstanding
B: 1.0 – 3.0 lbs	B: Good
C: 0.5 – 1.0 lbs	C: Okay
D: Less than 0.5 lbs	D: Nothing special
F: You don't show up to class	F: Divert your eyes!

1.6.5 Crash Test Dummy

Challenge: Design and build a LEGO car that can survive (not fall apart) a fall from waist high onto a hard floor.

Skill Badges: None

Procedures:

Experimental Setup: No specialized set up needed.

Robot Design: Your team must design and build a LEGO car that can survive a fall from waist high and not fall apart. For the purpose of this challenge, "waist high" is defined as the height of your waist. Here's one of those rare occasions where being short is an advantage!

Since creativity will be based on the appearance of your car you may want to add people (crash test dummies), headlights, and other amenities. Of course, these are the parts of the car most likely to fall off during the test. Engineering is all about making trade-offs, so you might as well get used to it.

Program: None

Hints: Bracing is the key to building strong LEGO structures. You should also get used to prototyping – testing early designs before the real test.

WARNING: DO NOT DROP AN RCX AND/OR MOTORS! You'll end up doing significant damage. The short stubby axle on the motor can be sheared off very easily and will render your $25 motor useless.

Grading:

Your grade will be based 80% on performance and 20% on creativity and aesthetics.

Performance	Creativity & Aesthetics
A: Survives the fall totally intact B: Nothing structural breaks off C: Car still rolls, despite missing some parts D: Hits the ground and shatters F: You don't show up to class	A+: Best of show A: Outstanding B: Good C: Okay D: Nothing special F: Divert your eyes!

1.6.6 Semester Clock

Challenge: Design and build a LEGO clock that counts down the days left in the semester.

Skill Badges: None

Procedures:

Experimental Setup: The instructor needs a LEGO motor and RCX programmed to turn the motor on at full power. Since the clock hand will be moving imperceptibly slow, the grade should be based on the gear ratio used.

Robot Design: Your team must design and build a LEGO clock that makes one full revolution in approximately 90 days. In order to accomplish this, you will need to calculate the gear ratio that results in one revolution in 90 days.

You MUST use the white clutch gear to limit the amount of torque. If you don't, we guarantee you'll end up with axles like the one shown if Figure 1.29!

The no-load speed of a LEGO motor is about 350 revolutions per minute. Since you will be using a lot gears, which can create a lot of friction, you can count on the LEGO motor spinning at about 200 RPM.

For this challenge, you will also have to make a clock face. Some plain paper with hand-written numbers and decorations will suffice.

Program: None

Grading:

Your grade will be based 80% on the gear ratio and 20% on creativity and aesthetics.

Gear Ratio	Creativity & Aesthetics
A: Gear ratio approximately right	A+: Best of show
B: Clock turns too slow to see	A: Outstanding
C: Clock turns very slowly	B: Good
D: Clock hands turn	C: Okay
F: You don't show up to class	D: Nothing special
	F: Divert your eyes!

- ½ grade if you don't use the white clutch gear

CHAPTER 2
GREEN LEVEL

Skill badges available in this Chapter

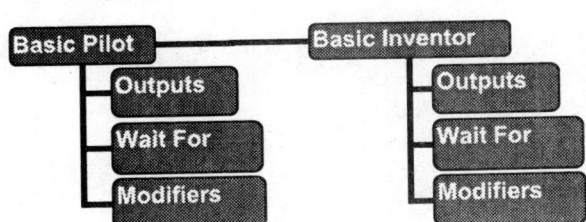

2.1 Green Challenges

The green challenges cover the basic pilot and basic inventor programming skills.

2.1.1 The Steepest Incline

Challenge: Using Pilot level 4, design and build a vehicle that is activated by a touch sensor and can climb the steepest incline.

Skill Badges: Basic Pilot

Procedures:

Experimental Setup: All you need for this challenge is an adjustable slope incline such as wide board and some bricks. Getting to the 'top of the hill' is defined as progressing up the slope a given distance (e.g. 12 inches) that will be specified by the instructor. Each attempt will last a maximum of 10 seconds.

Robot Design: Design and build a motorized vehicle using the LEGO RCX. The vehicle should start running once a touch sensor is pressed (by the instructor). The only construction restriction is that at most two motors can be used and that the vehicle must not leave anything behind.

Design factors to consider include gear ratios, friction, and center of mass of your robot.

Program: The program for this challenge is fairly simple; the program should wait for a touch sensor to be pressed before turning on the motor(s).

Hints: use the AC power adapter to prevent rapid use of batteries (stalling a motor will quickly drain a battery).

Grading:

Your grade will be based 75% on performance and 25% on creativity and aesthetics.

Performance	Creativity & Aesthetics
A+: The hill climb champion	A+: Best of show
A: Climbs at least a 30 degree incline	A: Outstanding
B: It moves uphill	B: Good
C: Motors turn on when touch sensor is pressed	C: Okay
D: You have something resembling a vehicle	D: Nothing special
F: You don't show up to class	F: Divert your eyes!

Like always, you must receive a "B-" or better to earn the Skill Badge.

2.1.2 Tug-of-War

Challenge: Design and build a tug-of-war robot using Pilot Level 4 to program the robot. The objective is to pull the opposing robot over the centerline. The battle commences when the robots are activated by a single touch sensor.

Skill Badges: `Basic Pilot`

Procedures:

Experimental Setup: The required materials for this Challenge include a line (electrical or masking tape) on the floor, two pieces of 6-inch long string, each with a paper clip at one end, and a touch sensor with two long lead wires.

During the competition, the opposing robots are tied together by the connecting the paper clips on the ends of the string. The paper clips are positioned over the line. The lead wires from the touch sensor are connected to Input Port 1 on both RCX's. The battle starts when the instructor presses the touch sensor (one touch sensor triggers both RCX's).

A single elimination contest will be used to determine which design wins. A draw will be declared after 10 seconds elapses without any progress.

Robot Design: The only three restrictions are:
1) Input Port 1 must be accessible to connect the touch sensor lead wire to.
2) You may use a maximum of two motors and
3) Your robot must permit a piece of string to be attached (you decide how).
4) Your robot must fit inside a cube 9 inches on a side.

Design factors to consider include gear ratios, friction, and center of mass of your robot.

Program: The program for this project is fairly simple. Your robot must wait until the touch sensor on Input Port 1 is pressed before turning on the motor(s).

Hints: Use the AC power adapter to prevent rapid use of batteries (stalling a motor will quickly drain a battery).

Grading:
Your grade will be based 75% on performance and 25% on creativity and aesthetics.

Performance	Creativity & Aesthetics
A+: Undefeated champion	A+: Best of show
A: You win more than once	A: Outstanding
B: You put up a good fight	B: Good
C: Motors turn on when touch sensor is pressed	C: Okay
D: You have something to connect the string to	D: Nothing special
F: You don't show up to class	F: Divert your eyes!

Like always, you must receive a "B-" or better to earn the Skill Badge.

2.1.3　Going the Distance

Challenge: Using Pilot mode, design, build, and calibrate a car that can travel a specified distance.

Skill Badges:　Basic Pilot

Procedures:

Experimental Setup:
　　In class: the instructor will need a line (electrical or masking tape), a tape measure, and an area clear of obstacles to run the robots. The instructor will also supply a computer with ROBOLAB installed to program the cars in class.
　　At home; you will need a tape measure to *calibrate* your car (distance vs. time). Make sure to ask the instructor what type of floor surface the Challenge will be conducted on in class (e.g. carpet or tile).

Robot Design: Design and build a motorized car using the LEGO RCX.

Program:
1. Program you car to travel for different amounts of time in PILOT (level 2 or higher) and record the distance it travels for a motor power level of 1. Depending on your design, the time will typically range from 0-5.5 seconds, which corresponds to approximately 0-10 feet.
2. **Calibration** of distance versus time: Using Excel, create a graph that shows how far your RCX car travels (in inches or centimeters) when programmed for a given amount of time.
   - The data collected should be shown as data points.
   - You should add Trendlines (linear regression) for each data set – make sure you display both the equation and r^2-value on the chart.

   Repeat steps 1 and 2 for motor power levels 3 and 5 and plot them on the same graph.

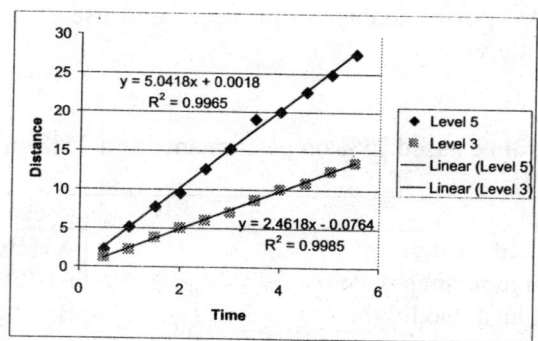

Sample calibration graph

3. Class Competition
   - Bring your car AND calibration graph to class.
   - The in-class competition will be to see who can get the closest to the line without crossing it.

- The distance to be traveled will not be revealed until class time (you should check with the instructor about the type of floor the competition will be held on).
- The instructor will provide a computer and ROBOLAB for programming your car in class.
- Your grade will depend on how close you get to the line.

Hints:
- Use the AC power adapter to make sure the results are not dependent on battery level.
- Think about how the speed of your car affects the accuracy of your calibration curve.
- Try to get an estimate of the repeatability and accuracy of your calibration curve. After you have created your calibration curve, try testing it out by using the regression equations to predict the distance traveled.

Grading:

Your grade will be based 80% on accuracy and 20% on creativity and aesthetics.

Accuracy	Creativity & Aesthetics
A+: You get the closest	A+: Best of show
A: You get within 6 inches	A: Outstanding
B: You stop short by more than 6 inches	B: Good
C: You go over the line	C: Okay
D: Your car runs when turned on	D: Nothing special
F: You don't show up to class	F: Divert your eyes!

Like always, you must receive a "B-" or better to earn the Skill Badge.

2.1.4 Tunnel Vision

Challenge: Using Pilot Level 4 or Inventor Level 4, design, build, and program a vehicle that automatically turns its headlights on when it enters a dark tunnel. You can use either Inventor or Pilot mode, which will determine the Skill Badge you earn.

Skill Badges: Basic Pilot OR Basic Inventor

Procedures:

Experimental Setup: This Challenge requires a tunnel of sufficient size for a robot/vehicle to drive into. Lamp elements (see Table 1.1) are needed to serve as the headlights. If no Lamp elements are available, a sound can be played instead. Note, the sound output is not available in Pilot mode.

Robot Design: Design and build a motorized car using the LEGO RCX that has a light sensor on it.

Program: Your robot should move forward when turned on. When the light sensor detects dark, you should turn the lamp element on (or play a sound if no lamp elements are available) while continuing to move forward (i.e. your car should not have to stop to turn its lights on). When it emerges from the other end of the tunnel, the lamp should turn off.

Hint: The amount of ambient room light can make a big difference when using light sensors. You might want to practice in the classroom to determine the proper light sensor threshold setting. Also, if you are using Inventor Mode, you will have to choose between the ***wait for dark*** and ***wait for darker*** commands.

Grading:

Your grade will be based 75% on performance and 25% on creativity and aesthetics.

Accuracy	Creativity & Aesthetics
A: Turns lights on & off properly	A+: Best of show
B: Turns lights on when entering the tunnel	A: Outstanding
C: The lights go on and off (anytime)	B: Good
D: Your car runs when turned on	C: Okay
F: You don't show up to class	D: Nothing special
	F: Divert your eyes!

Like always, you must receive a "B-" or better to earn the Skill Badge.

2.1.5 Wallace & Gromit™

Challenge: Design and build a device that can deliver a person (LEGO mini-fig) from a bed on the 2nd floor to the seat at the kitchen table on the 1st floor. The system should be activated by the rising sun (simulated by a flashlight). You can use either Inventor or Pilot mode, which will determine the Skill Badge you earn.

Skill Badges: Basic Pilot   OR   Basic Inventor

Procedures:

Experimental Setup: We highly suggested watching the Wallace & Gromit animated feature *The Wrong Trousers* as an introduction to this Challenge. Both a flashlight and the first floor of the house, including the table and chair, will be supplied by the instructor.

Robot Design: The only restriction is that the device used to transport the mini-fig must not be touching the mini-fig at the end of the challenge (i.e. you must let go at the end). You must build the second floor of the house, including the bed. This Challenge is more build-intensive than program-intensive.

Program: Your program must be activated by a light sensor. The program will most likely consist of turning motors on and off at specified times.

Grading:

Your grade will be based 70% on performance and 30% on creativity and aesthetics.

Accuracy	Creativity & Aesthetics
A: In the chair, not touching the floor B: Touching the chair and/or table C: Somewhere on the first floor D: In bed upstairs F: You don't show up to class	A+: Best of show A: Outstanding B: Good C: Okay D: Nothing special F: Divert your eyes!

Like always, you must receive a "B-" or better to earn the Skill Badge.

The Wallace and Gromit characters are trademarked and copyrighted by Aardman Animations

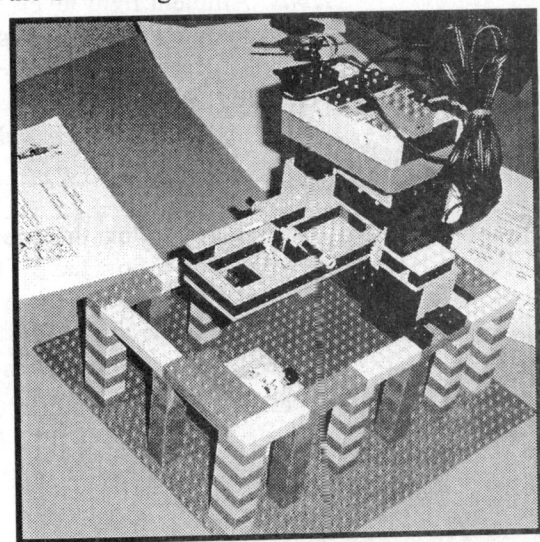

2.1.6 Line Follower

Challenge: Design and build a robot that can follow a black line. You can use either Inventor or Pilot mode, which will determine the Skill Badge you earn.

Skill Badges: [Basic Pilot] OR [Basic Inventor]

Procedures:

Experimental Setup: The course should be constructed using black electrical tape on a white background. The line will curve gently and be approximately 3 feet long. The instructor will use a stop watch to time how long it takes your robot to follow the line.

Robot Design: For this exercise you need to design and construct a basic car with at least two wheels and a light sensor facing down.

Program: There is a very simple **algorithm** that will make a robot follow a line. Actually, this **algorithm** doesn't really follow a line as much as zigzag back and forth over the line. Assuming you start on the line:

```
Turn left motor on
Wait for bright
Turn left motor off
Turn right motor on
Wait for dark
Turn right motor off
Repeat
```

> An **algorithm** is an outline of the major steps required to solve the problem

By alternating the left and right motors, the robot will pivot about the wheel which is stopped and slowly crawl forward on the right side of the line. While this algorithm is very robust (works nearly every time), you will quickly find that this method of line following is pretty slow. To speed up your robot you can leave the motor on at a low power level rather than turn it off completely. If you set the motor power too high, the robot will tend to loose the line. You should try experimenting with the motor power levels to get a good combination of reliability and speed.

Hints: Ambient light levels can drastically affect the performance of the line follower. Try shielding the light sensor from all ambient light to increase reliability (see Fig. 2.12).

Grading:

Your grade will be based 75% on performance and 25% on creativity and aesthetics.

Accuracy	Creativity & Aesthetics
A+: You follow the line the fastest	A+: Best of show
A: You follow the entire line	A: Outstanding
B: You follow the line more than 1 foot	B: Good
C: You car zig-zags	C: Okay
D: Your car runs when turned on	D: Nothing special
F: You don't show up to class	F: Divert your eyes!

2.1.7 Speed Walking

Challenge: Design and build the fastest walking robot you can. Just to make things interesting, you will be racing over a pebble surface. You can use either Inventor or Pilot mode, which will determine the Skill Badge you earn.

Skill Badges: Basic Pilot   OR   Basic Inventor

Procedures:

Experimental Setup: This challenge is typically run over a pebble course in a single elimination format with the winner of each race advancing to the next round. Typically the race course is about 4 feet long.

Robot Design: For this exercise you need to design and construct a robot that can walk (crawling may also be permitted) over a mildly uneven surface. Creativity is worth 75% and will be based largely on the walking/crawling mechanism employed.

The only restriction is that you cannot intentionally trip another robot.

Program: Most likely your program will simply consist of turning motors on and off. This a build-intensive, not program-intensive, challenge.

Grading:
Your grade will be based 25% on performance and 75% on creativity and aesthetics.

Performance	Creativity & Aesthetics
A+: You are the fastest walker	A+: Best of show
A: You walk and win at least once	A: Outstanding
B: You walk, but don't win	B: Good
C: You walk but don't make it to the finish line	C: Okay
D: You move, but don't walk	D: Nothing special
F: You don't show up to class	F: Divert your eyes!

2.1.8 How Fast is That?

Challenge: Design and build a robot to measure the torque versus RPM of a LEGO motor.

Skill Badges: [Basic Inventor]

Procedures:

Experimental Setup: This Challenge requires the use of a LEGO rotation sensor (see Table 1.2). The instructor will supply calibrated weights for you to use. You will also need a timing device such as a stop watch. This Challenge is done completely as homework.

Robot Design: For this exercise you need to design and construct a robot that can wind up a string with a weight on the end. Using a stop watch, you will measure the length of time it takes to reach the specified number of rotation units – this will allow you calculate the approximate RPM (revolutions per minute) of the LEGO motor. By varying the weight on the string and/or the diameter of the hub you wind the string on, you can vary the torque applied to the LEGO motor. You should acquire approximate 15-20 data points.

You must submit a copy of your program, the torque-RPM curve, and a digital photograph of the robot used.

Program: You should program your robot to turn on a motor for a specified number of rotation units (16 rotation units = 360 degrees).

Bonus (extra half-grade): Most electric motors exhibit **asymmetry**, meaning they have different characteristics in the forward and reverse directions. Measure the torque-RPM in both directions for an extra half-grade.

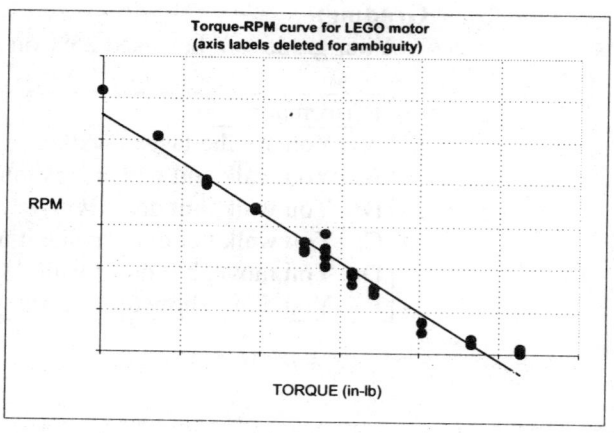

Grading:

Your grade will be based 90% on performance and 10% on creativity and aesthetics.

Accuracy	Creativity & Aesthetics
A: You submit all 3 items.	A+: Best of show
B: You submit 2 of the 3 items.	A: Outstanding
C: You submit only one of the 3 items: torque-rpm curve, program or digital photograph.	B: Good
	C: Okay
D: You submit documentation of a valiant attempt to complete the Challenge	D: Nothing special
	F: Divert your eyes!
F: You submit someone else's graph, program or digital photograph	

Like always, you must receive a "B-" or better to earn the Skill Badge.

2.2 Pilot Basics

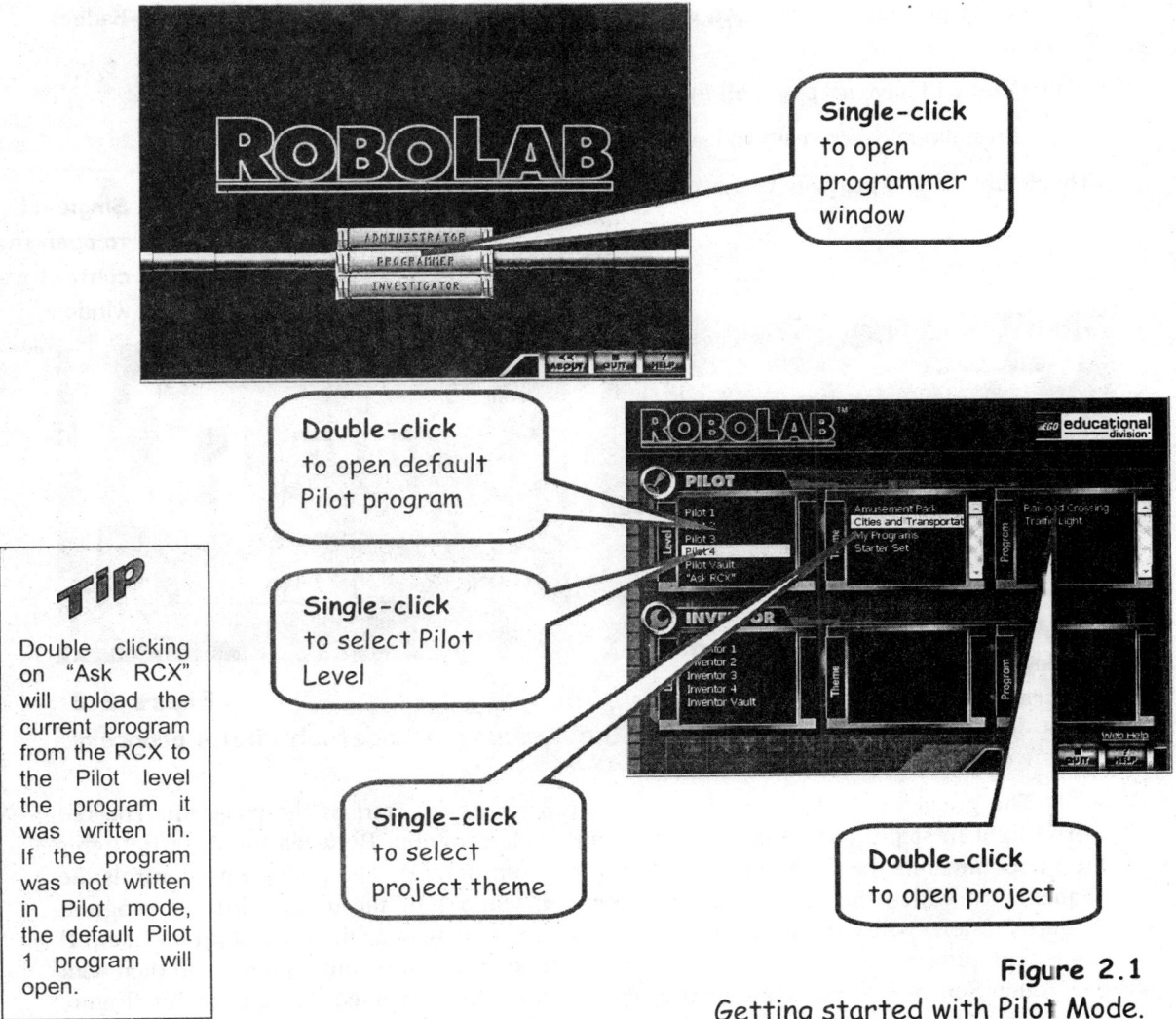

Figure 2.1
Getting started with Pilot Mode.

TIP

Double clicking on "Ask RCX" will upload the current program from the RCX to the Pilot level the program it was written in. If the program was not written in Pilot mode, the default Pilot 1 program will open.

Pilot mode is the easiest of the 3 programming modes in ROBOLAB. Pilot mode can be accessed from the Programmer window. Pilot mode is very useful for getting familiar with the RCX, outputs, and sensors because all Pilot mode programs will *always* execute. They may not do exactly what you intended, but they will always compile and execute.

Pilot programs are sequential, meaning the commands are executed one after another in a fixed sequence. If you are familiar with computer programming already, you will notice that loops, functions (subroutines), and other programming structures do not exist in Pilot mode.

Pilot Mode has four levels, with Pilot 1 being the simplest and Pilot 4 the most complex. Pilot 3 and 4 each have several Themes with sample Programs.

2.3 The Basic Pilot Badge

This section provides information on the skills needed to earn the Basic Pilot skill badge. Rather than start with Pilot 1, we're going to get started with Pilot 4 for two reasons:

1) You will have access to all the functions and
2) Pilot mode is very easy to learn (we have full confidence in you).

The default Pilot 4 program is shown below in Figure 2.2.

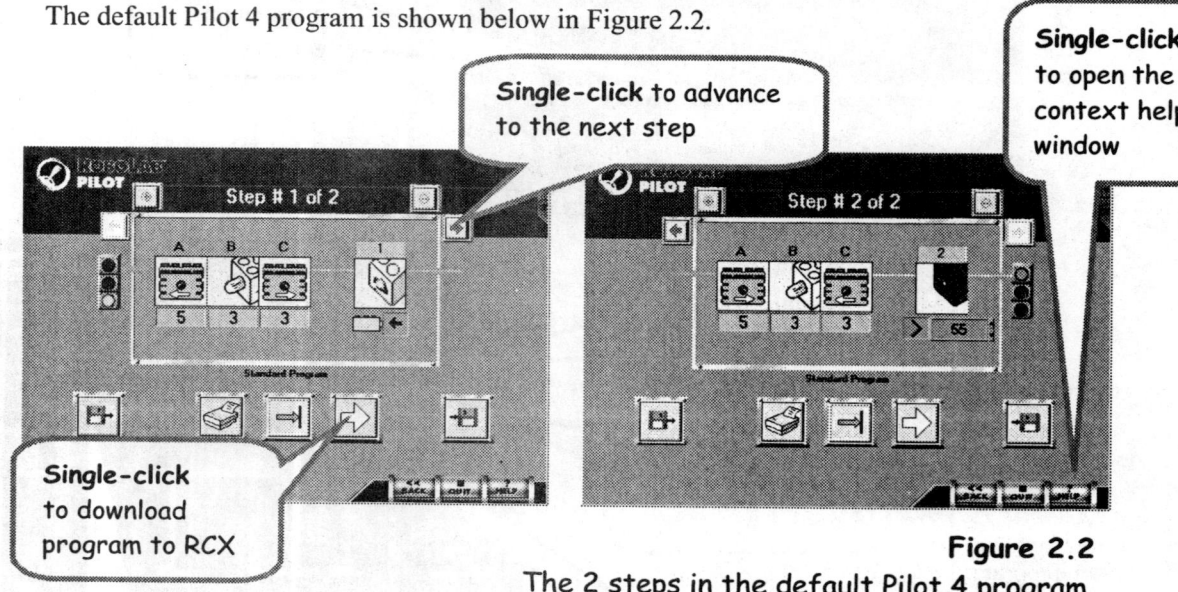

Figure 2.2
The 2 steps in the default Pilot 4 program.

The green traffic light in step 1 (left) represents the start of the program. The red traffic light in step 2 (right) indicates the end of the program. Between the 2 traffic lights are the commands the RCX will carry out (i.e. the program). The pink string controls the sequence in which the commands are executed. Step 1 of the default Pilot 4 program (Figure 2.2-left) will turn on Motor A at full power in the reverse direction, turn on Lamp B at medium power, turn on Motor C in the forward direction at medium power and then *wait for* Touch Sensor 1 to be pressed. After the touch sensor is pressed step 2 executes (Figure 2.2-right): turn on Motor A in the forward direction at full power, turn on Lamp B at medium power, turn on Motor C in the reverse direction at medium power, *wait for* the reading from Light Sensor 2 to exceed 55 and then stop.

Clicking the large white arrow icon will download the program to the RCX. Make sure the RCX is turned on and the IR sensor on the front RCX is directly facing the IR transmitting tower attached to your computer. Use the **Prgm** button (gray) to select which of the 5 program slots to use. A task meter will indicate download progress (Figure 2.3 - left).

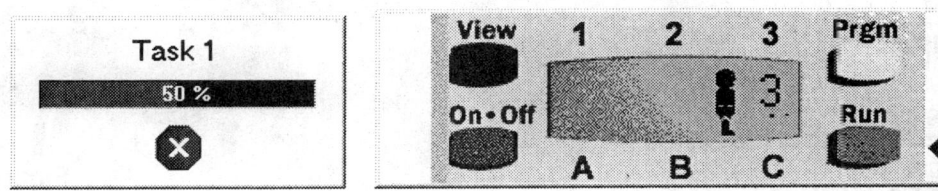

Figure 2.3
Download progress (left) and *download complete* (right) indicators.

Once the download is complete, a window indicating the program slot used will appear (Figure 2.3 – right). Note that if you attempt to download to program slots 1 and 2 while they are locked, program slot 3 will be used. See Chapter 1.3.1 for information on how to unlock program slots 1 and 2.

2.3.1 Outputs

Clicking on a motor or lamp icon opens a pop-up window which allows you to select one of four output commands: *motor reverse*, *motor forward*, *lamp on*, or *stop output*. The *stop output* command is used when you either want to turn a motor or lamp off or when there is no output device connected to a port. The power level is shown below the motor/light icons and will be discussed later in Section 2.3.3.

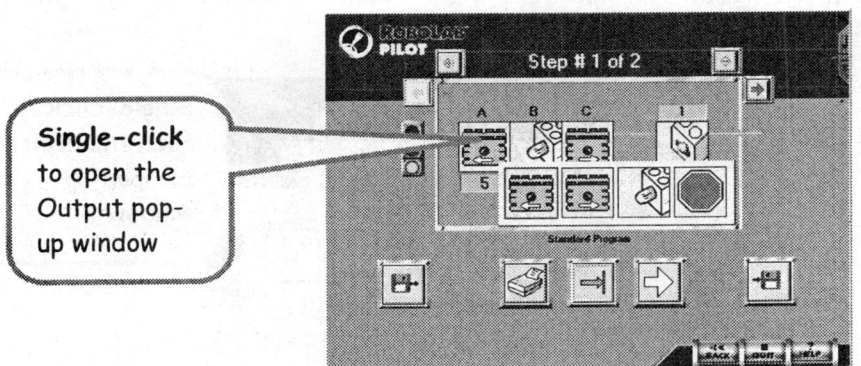

Figure 2.4
Output selector pop-up window.

While the commands are called *motor forward* and *motor reverse*, the actual direction the motor spins will depend on the orientation of the lead wires. Since there are 4 orientations a lead wire can be attached and there are connections at both the motor and the RCX, there are 16 possible ways to connect a motor to the RCX. Fortunately, many of the orientations have the same result. Shown below are pairs of connections that cause the motor to rotate in the same direction. In general rotating a connector 180 degrees will reverse the motor direction.

TIP

Rotating the lead wire 180 degrees at either the motor or the RCX will always flip the direction of rotation.

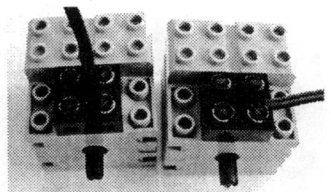
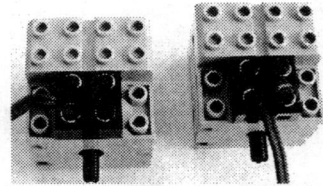

Figure 2.5
Motor connections that result in the same output.

Figure 2.6
RCX connections that result in the same output.

Finally, while only the geared motor and lamp element are listed as output devices in ROBOLAB there are actually several useful LEGO output devices that you can buy. A complete list of output devices is shown in Table 1.1 in Chapter 1.

2.3.2 The "Wait For" Functions

The *wait for* command is probably the most commonly used command in ROBOLAB. Clicking on the touch sensor icon opens a pop-up window (Figure 2.7) where you can select one of three *wait for* conditions: time, touch, or light.

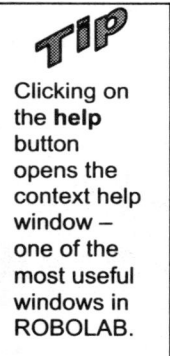

TIP

Clicking on the **help** button opens the context help window — one of the most useful windows in ROBOLAB.

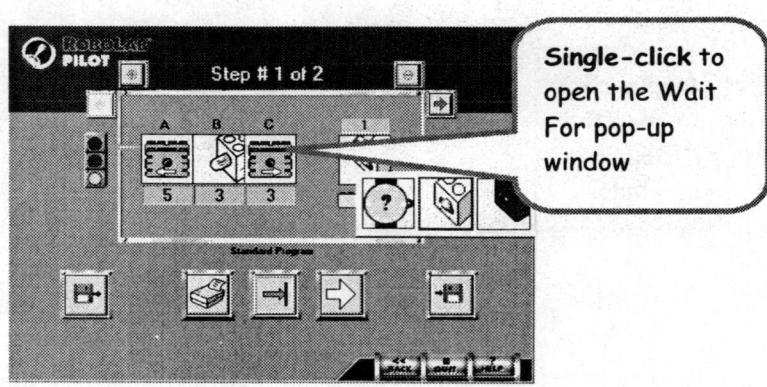

Figure 2.7
Wait for selector pop-up window.

As shown in Figure 2.8, the *wait for time* command can either be set for a specified time or a random time (between zero and specified maximum). The *wait for touch* command can be set to wait for either a press or a release. Finally, the *wait for light value* command can be set for wait for either dark or light (100 = bright, 0 = dark).

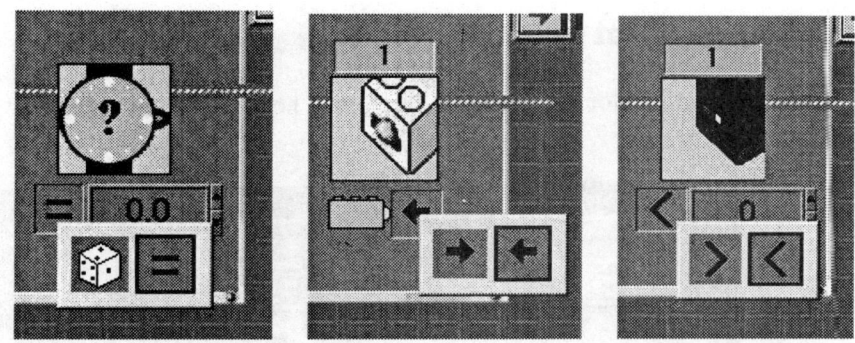

Figure 2.8
Pilot 4 *wait for* modifiers.

2.3.3 Modifiers

Modifiers allow you to select which port a sensor is connected to or at what power level to run a motor (see Figure 2.9). Modifiers also allow you specify the *wait for* conditions like length of time or light threshold value (see Figure 2.8).

In Pilot level programming the *power level modifiers* are located below the motor and light icons. They indicate the power level being supplied to each of the output ports. Clicking on the modifier allows you to change the power level of each port. Level 5 is the highest level (brightest light or fastest motor) and Level 1 is the lowest level.

Similarly, the number above the touch and light sensor *wait for* command is the *input port modifier*. Clicking on it allows you to specify the input port to which the sensor is connected.

The most common mistake for beginners is to connect a sensor to the wrong input port. For example, Input Port 1 is specified in the program, but the sensor is connected to Input Port 2 on the RCX.

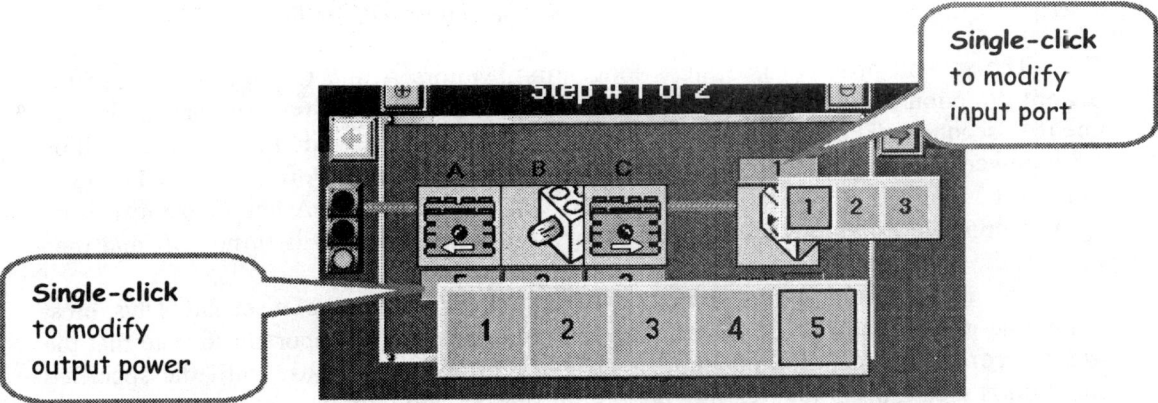

Figure 2.9
Power level and input port modifiers.

Strictly speaking modifiers adjust motor power not motor speed. While more power can lead to more speed, if the robot is extremely overpowered, adding more power will not affect speed very much. Thus, power level 1 does not always result in a slower robot than power level 5 (see Chapter 4.7 for more details on power versus speed)

2.3.4 Sample Pilot Level 4 Programs

Here are two sample Pilot 4 programs which will help you understand how to create your own Pilot programs.

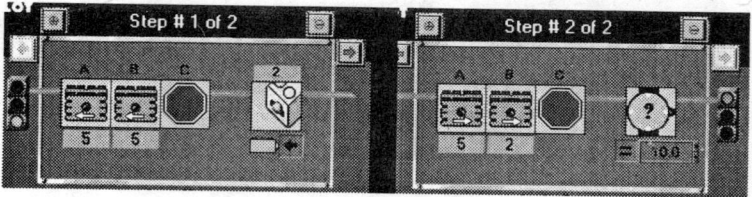

Figure 2.10
Sample Pilot program.

This first program is two steps long. The program starts at the green light (far left) and then turns on both Motors A and B in the reverse direction at full power (level 5) and then waits for the touch sensor on Input Port 2 to be pressed. Then Motor A is turned on in the forward direction at full power and Motor B is turned on in the forward direction at power level 2 and then waits for 10.0 seconds before stopping at the red stop light (far right).

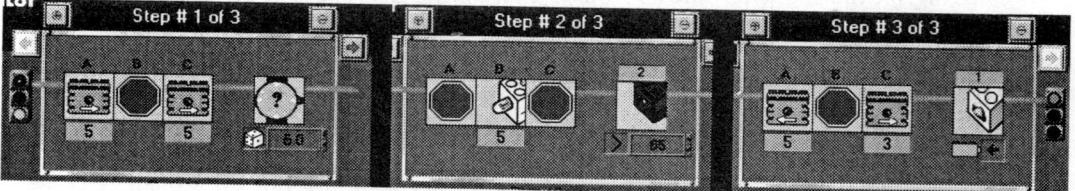

Figure 2.11
Sample pilot program with three steps.

The second program is 3 steps long. First, Motors A and C are turned on in the forward direction at full power (level 5) and then the program waits a random amount of time (0-5 seconds). Then in step 2, Motors A and C are stopped while Lamp B is turned on at full power. The program then waits for the light sensor value on Port 2 to exceed 65 (100 = bright, 0 = dark) before proceeding to step 3. In step 3, Motors A and C are turned on again in opposite directions (A reverse, C forward) while Lamp B is turned off and then waits for the touch sensor on Port 1 to be pressed before stopping.

The actual turning of motors on and off only takes a fraction of a second. Thus, most of the time in a program is spent at the *wait for* commands. It is important to note that the *wait for* command does nothing more than make the program pause until the specified condition is met. During this waiting period, all sensors and outputs will be active; motors will continue to run and sensors will continue to function.

2.3.5 Notes About Using the LEGO Light Sensor

Not surprisingly, the light sensor works by measuring the intensity of light (in both the visible and infrared wavelengths). The light source can either be external (e.g. a flashlight) or internal (i.e. the red LED). In practice, you will use the light sensor in one of two ways: to measure either the amount of light reflected by a surface (reflective mode) or the intensity of an external light source (ambient mode). Probably the two most common uses for the light sensor each use one of the two modes: line following (reflective mode) and finding the brightest spot in a room (ambient mode).

REFLECTIVE MODE: measuring the amount of light reflected by a surface utilizes the internal red LED. In this measurement mode, the red LED emits light and it bounces off a surface and the amount of reflected light is read by the light sensor. The amount of light reflected by the surface will depend on lots of things. For example, a rough white surface far away can result in the same amount of reflected light as a smooth black surface close up.

We've seen many students yelling at a computer because they can't get the light sensor to reliably sense color during a line following exercise. The root of their frustration is that the light sensor doesn't measure color – it measures the amount of light. Color, surface roughness, distance to the surface, angle of incidence, and amount of ambient light will all affect the amount of light reflected. Black will reflect less light than white. A rough surface will reflect less light than a smooth one. Surfaces close up will reflect more light than surfaces far away. Surfaces perpendicular to the light sensor will reflect more light than those at an inclined angle. And finally, ambient light (both visible and infrared) is the largest source of error (you measure it, but don't want to).

If this sounds complicated, it's because it is. However, it's the large number of contributing factors that also makes the light sensor so versatile. In practice, you should try to measure the effect of just one factor, such as color, while keeping all other factors constant. For example, if you want to sort LEGO bricks by color, make sure all the bricks are the same distance from the light sensor and that they are all presented to the light sensor at the same angle and there is no ambient light.

On the other hand, if you want to measure the distance to a surface (a proximity sensor), along with minimizing ambient light you should make sure the surface is perpendicular to the light sensor, and that color and surface roughness are uniform.

In general, we can use the light sensor to measure any of the factors mentioned above: color (color sensor), surface roughness (roughness meter), distance (proximity detector), and angle of incidence (inclinometer). The only caveat is that we minimize the amount of ambient light because ambient light is the greatest source of error.

If the amount of ambient light was constant, it would simply cause our measurement of reflected light to be a little too high. However, in reality the amount of ambient light changes drastically as your robot moves around in and out of shadows. The effect of changing ambient light levels is usually larger than the effect of the factor you are trying to measure.

If you could use the light sensor in complete darkness, the repeatability of the readings would be quite high. This is the ideal condition, but isn't very practical. To simulate complete darkness you can place the light sensor under the robot where it is shaded from most external light sources. In this way, it wouldn't matter if the room lights were on or off – the robot will be using the red LED as the only light source. You can also isolate the light sensor as shown in Figure 2.12. This doesn't block all ambient light, but helps quite a bit. Even better is the light sensor "sled" shown in Figure 2.13. Not as versatile as the using the 1x2 brick, but it does a much better job of blocking out all ambient light.

The light sensor has both a light source (red LED) and a sensing unit on it. The best way to make sure that your light sensor readings are consistent is to make the LEGO light source the only light source. This can be accomplished by isolating the light sensor from any ambient light. One creative way to is by using a 1x2 technics beam placed at the end of the light sensor as shown.

Figure 2.12
Isolating a light sensor from ambient light.

Figure 2.13. Light sensor "sled."
A light sensor "sled" that can be pulled behind or pushed in front of a robot and does a good job at isolating the light sensor from nearly all the ambient light.

AMBIENT MODE: measuring the intensity of an external light source is another common use for the LEGO light sensor. However, it is also much harder to do reliably than most people realize because of two things: the red LED and infrared (IR) sources.

Because the red LED is so close to the light detecting element (a phototransistor), it can make measuring weak external light sources nearly impossible. Even in complete darkness the light sensor will read around 20. Put your finger over the light sensor in an attempt to cover it up and you'll get a reading around 45. Any objects within a few inches will also cause the LED's reflected light to become significant.

In addition to the LED, light sensor readings can go awry due to interference from IR sources. The LEGO light sensor is actually very good at measuring the intensity of IR light sources which, unfortunately, we can't see. And the news gets worse, because IR sources are everywhere: light bulbs (incandescent and florescent), remote controls, heat sources, handheld computers, and the IR tower used to program the RCX.

Fortunately, the most common light source people try to detect is a standard flashlight, which happens to be both very bright in comparison to the surroundings and puts out a lot IR as well as visible light.

 Don't forget you can use the **view** button on the RCX to see what the current light sensor reading is.

Figure 2.14. Light sensor flipper.

One challenge that students often face is how to use the light sensor to both find an object (or an opponent) and keep and eye on the ground (to see the tournament boundaries). One solution is to use a differential to flip the light sensor up and down. In the photos shown, when moving forward the light sensor flips down. When reversing, the light sensor flips up to a horizontal position (not shown). Black friction pegs serve as positive stops.

2.3.6 Steps, Run Mode, Printing, and Saving

So what are all those other buttons on the Pilot screen? In case you haven't figured it out by playing with them, let's briefly go over the functions of the various buttons.

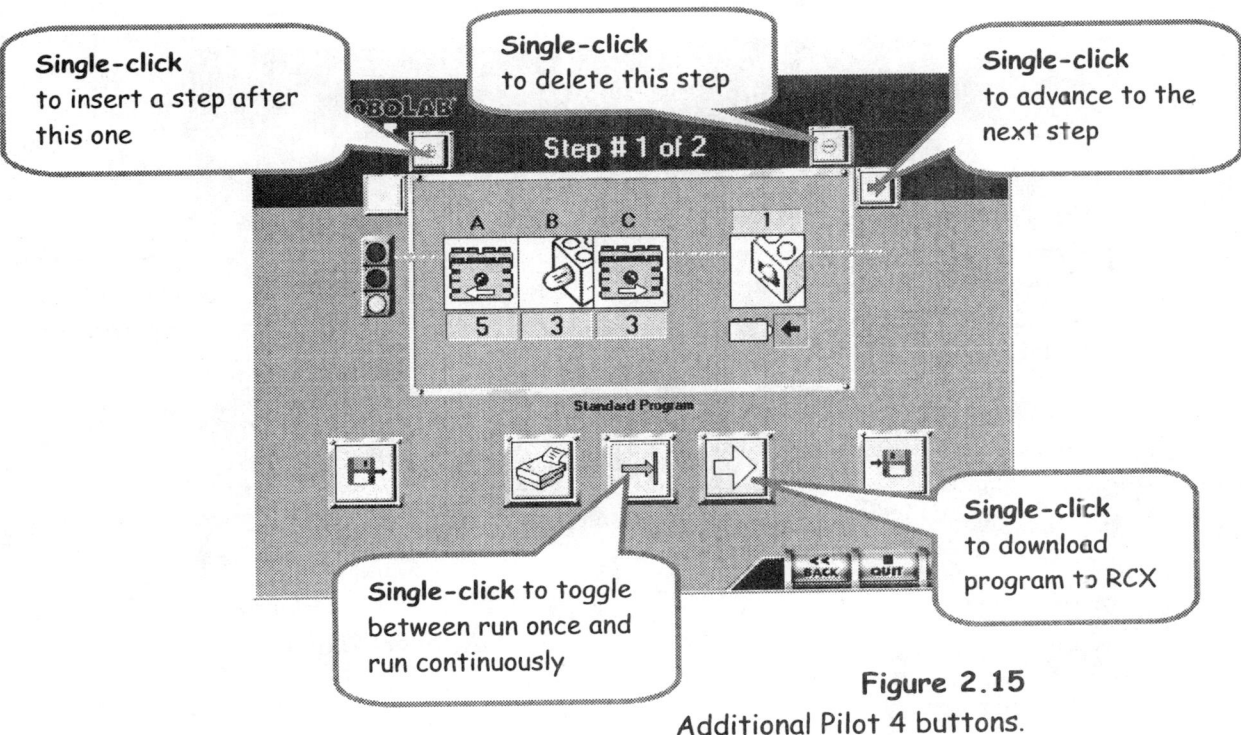

Figure 2.15 Additional Pilot 4 buttons.

Steps

Program steps are like frames in a movie. Each step is executed sequentially one after the other to make up the entire program. Pilot 4 allows you to create longer programs by adding steps. You insert or delete steps using the "+" and "-" icons at the top of the screen. You scroll backwards and forwards through the steps by using the red arrow icons on the top left and right of the Pilot window.

The sample program in Figure 2.10 has 2 steps. The sample program in Figure 2.11 has 3 steps. The maximum number of steps is unlimited.

Run Mode

The run mode button (pink arrow) toggles between *run once* and *run continuously*. If you select the *run continuously* option and run your program on your RCX, your program will start over and repeat until you stop it by pressing the green **run** button on the RCX again. This can be very useful for making a robot repeat a simple behavior.

Figure 2.16
Run mode.

If run continuously, the program in Figure 2.10 could be used to make a robot avoid obstacles. The robot would drive forward until it bumped into something, backs up and then goes forward until it bumps into something again. Because the motors are at different power levels in Step 2, the robot would tend to turn as it backed up.

Once you've selected the run mode, you still must download your program to the RCX using the large white arrow (run icon).

Printing & Saving

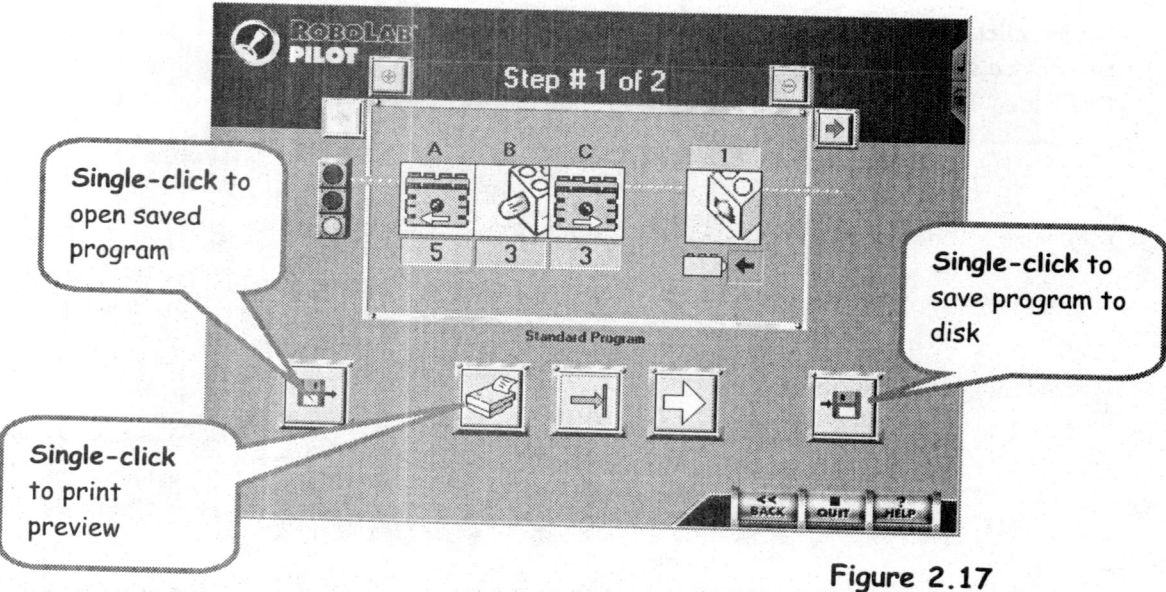

Figure 2.17
Additional Pilot 4 buttons.

Clicking on the printer icon in Figure 2.17 will open the print-preview window shown in Figure 2.18. Click on the **Print** icon to print the image or click on the **Back** button to cancel the print job.

Clicking on the **Save Program** button (lower right in Figure 2.17) will allow you to save your Pilot program to disk. Pilot Level 4 programs are saved as *filename.pi4*.
Similarly, the **Load Program** button allows you open previously saved Pilot programs.

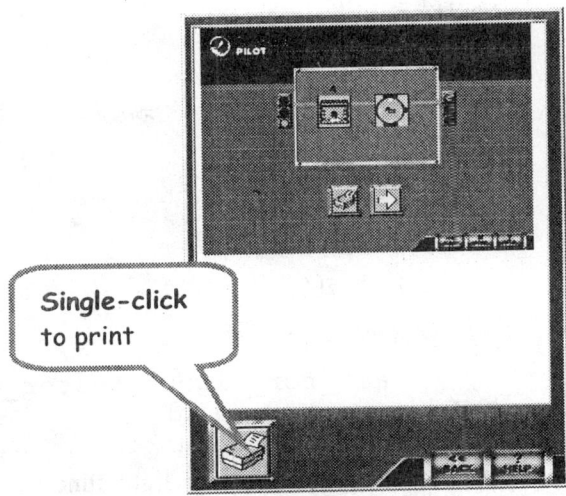

Figure 2.18
Print preview window.

2.4 Relation to Text-Based Programming

The icons in ROBOLAB correspond to *functions* in a text-based programming language and modifiers correspond to *arguments*. Figure 2.19 shows a 2 step Pilot Level 4 program.

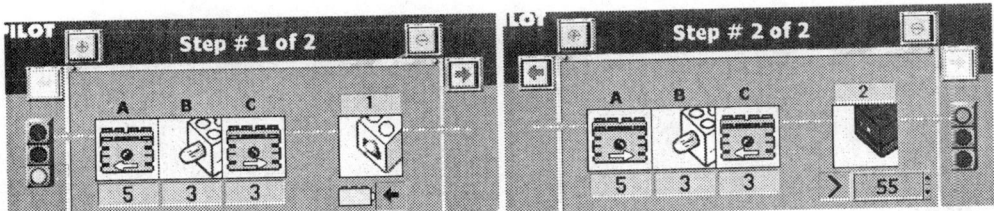

Figure 2.19
Sample Pilot program.

In step 1 the program begins by turning on Motor A at full power in the reverse direction, Lamp B at medium power, and Motor C at medium power in the forward direction. The program then waits (with motors and lights on) until the touch sensor connected to input port 1 is pressed. The program then proceeds to step 2, which starts by turning on Motor A in the forward direction at full power, Lamp B at medium power, and Motor C in the reverse direction at medium power. The program then waits until the light sensor connected to input port 2 registers a light level greater than 55 (100 = bright, 0 = dark) before stopping.
If we were to convert this program into a text-based programming language, it might look something like this:

```
Start;
    Rev_A(5);
    Fwd_B(3);
    Fwd_C(3);
    Wait_for_touch(press);
    Fwd_A(5);
    Fwd_B(3);
    Rev_C(3);
    Wait_for_light(greater_than,55);
End;
```

Note that the wait for light function is the only function that has two arguments, which correspond to the two modifiers in ROBOLAB (greater/less than and the threshold level).

For the novice programmer the main advantage of using ROBOLAB over a text-based programming language (such as NQC) is that you don't have to deal with syntax errors. Instead of worrying about spelling errors ("Rew_B" instead of "Rev_B") or which comes first in the argument list ("greater_than" or "55") you can focus on the logic of the program. You also don't have to worry about whether "55" is an integer, a floating point number, or a character string – you just use it.

In case you're interested, ROBOLAB actually converts the icon program you write into a text based program for use in LASM, the LEGO Assembly language, which is covered in Chapter 6. ROBOLAB does the conversion for you so that you don't have to worry about syntax.

2.5 Limitations of Pilot Programming

The great strength of Pilot Mode is that all programs will run (compile and execute). However, it's the sequential organization of Pilot programs, which makes it impossible to write a program that won't run, that is also the main limitation of Pilot mode.

Pilot mode does not contain any program *control structures*. In computer programming lingo, a *control structure* is a command which allows looping and/or branching in the program. In other words, *control structures* allow you to jump around in the program in a non-sequential manner. For example, if you wanted a robot wait for 2 seconds if touch sensor 1 is released (not pressed) and run Motor A for 10 seconds if touch sensor 1 is pressed, you couldn't do this in Pilot mode because it requires a *conditional statement*, which does not exist in Pilot Mode..

We need to step up to Inventor Mode to access conditionals and other advanced *control structures*. Shown below is a non-sequential program that would accomplish this simple task. Notice that the *arguments* are now "wired" to each *function*.

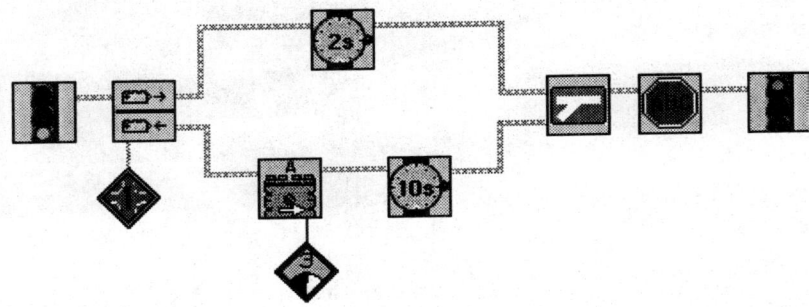

Figure 2.20
Sample conditional statement implemented in Inventor Mode.

In a text-based programming language, this program might look something like this:

```
Start;
    If_touch(1) = release then;
        Wait_for_seconds(2);
    If_touch(1) = press then;
        Fwd_A(3);
        Wait_for_seconds(10);
    End_if;
    Stop_ABC;
End;
```

2.6 Inventor Basics

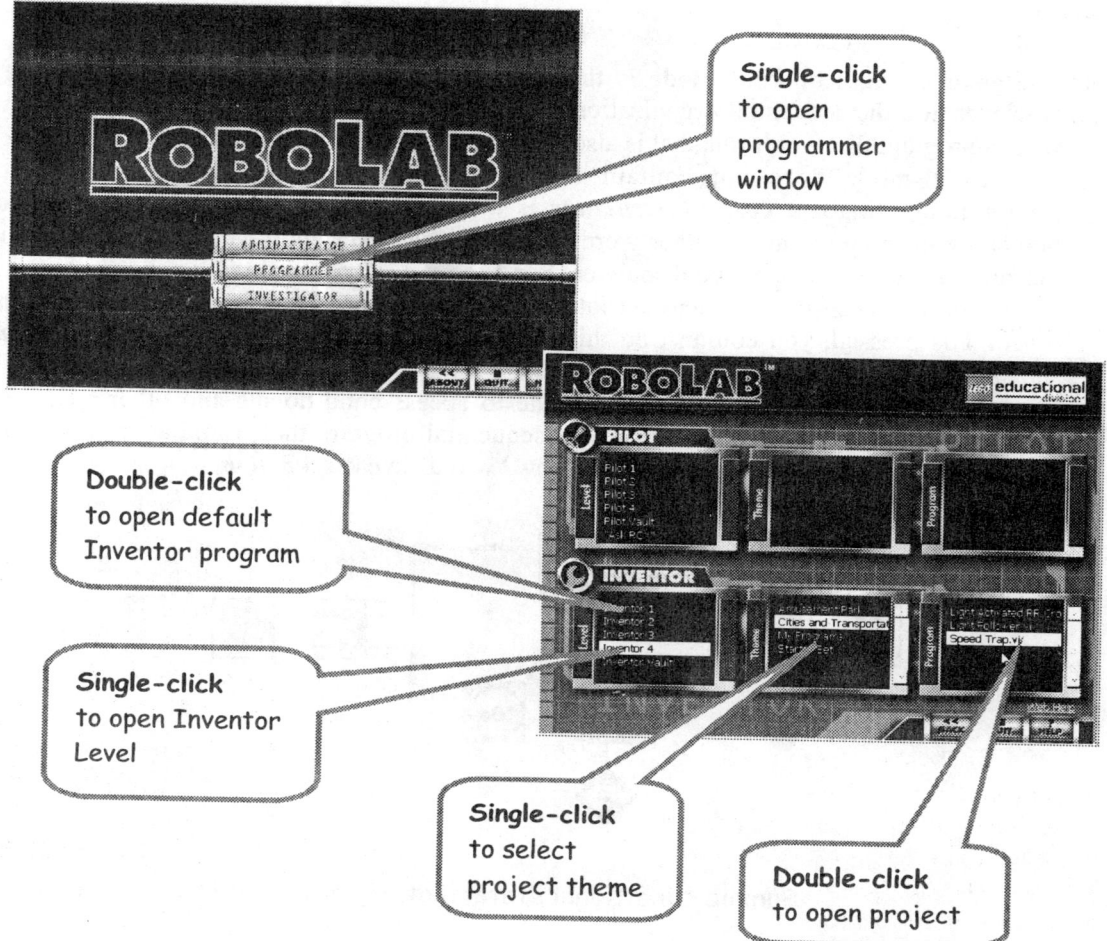

Figure 2.21
Getting started with Inventor Mode.

Inventor mode is the second of the three programming modes in ROBOLAB. Like Pilot mode, Inventor mode is accessed from the Programmer window (see Figure 2.21). Also just like Pilot mode, there are 4 levels of Inventor mode with Inventor 1 being the simplest and Inventor 4 being the most powerful.

The major differences between Pilot and Inventor modes are that you can make use of variables, structures, and subroutines in Inventor mode. Additionally, in Inventor mode you have to "wire" the icons together yourself. These additional features add complexity, but it also allows you to create very powerful programs.

This section will cover a few of the basics needed to get around in Inventor mode: the windows, tools palette, and function palette. The next section will cover the nuts and bolts of programming with Inventor.

When you open an Inventor program, three windows open. The upper window is the **Panel window**, the lower window is the **Diagram window**, and the floating window is the **Functions Palette**. All your programming will be done in the **Diagram window**.

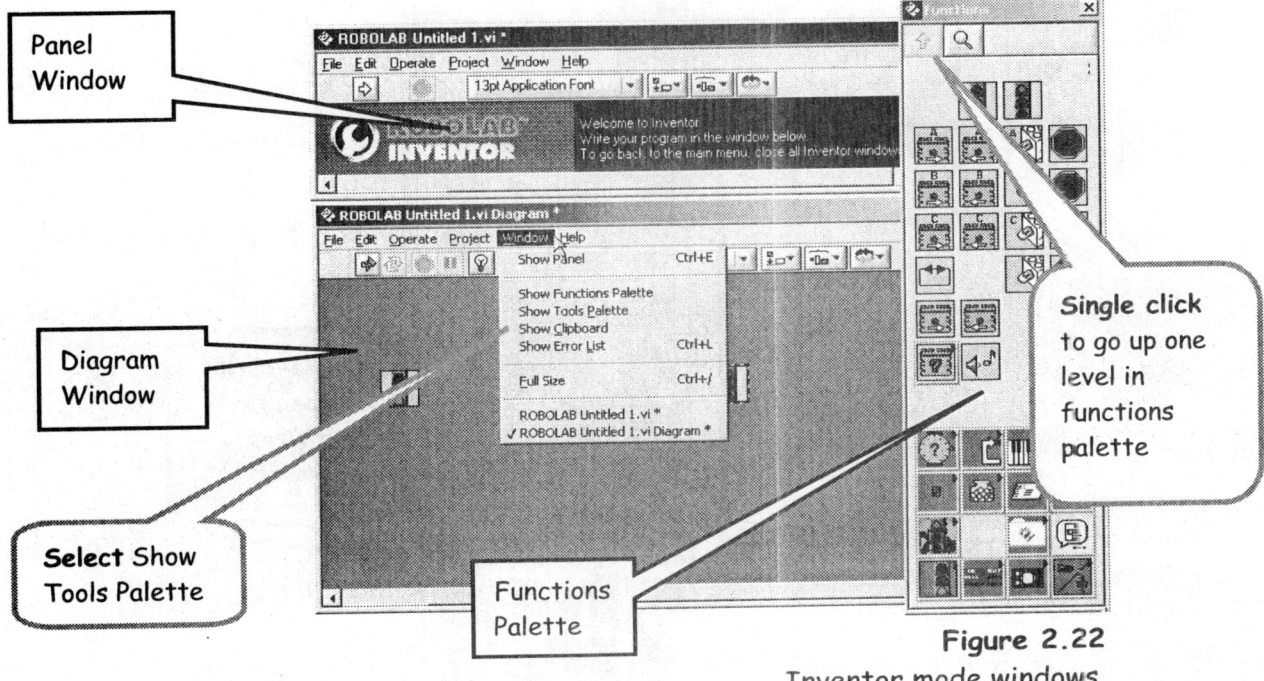

Figure 2.22
Inventor mode windows.

Remember the big white arrow in Pilot that was used to send your programs to the RCX? Well, its still here in Inventor, its just not as big anymore. You can find the **Run** button at the upper left hand corner of the **diagram window.**

The run continuously will NOT cause the program to restart. Instead it will cause the program to download to the RCX over and over and over.... (not a nice thing to do).

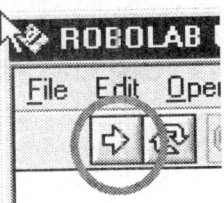

Figure 2.23
Run button.

2.6.1 The Functions Palette

The **functions palette** contains the programming icons used in ROBOLAB. Icons are picked from the **functions palette** and placed into the **diagram window**.

At the top of the functions palette are the green and red traffic lights (the begin and end icons). Just like in Pilot mode, every program must start with the green traffic light and end with the red traffic light.

The upper portion of the **functions palette**, as shown in Figure 2.24, contains all the output functions, most of which should look familiar from Pilot mode. The lower portion of the palette contains several sub-menus, which have the small black triange in the upper right corner of the icon. In this chapter, we will discuss the *wait for* sub-menu and the *modifiers* sub-menu.

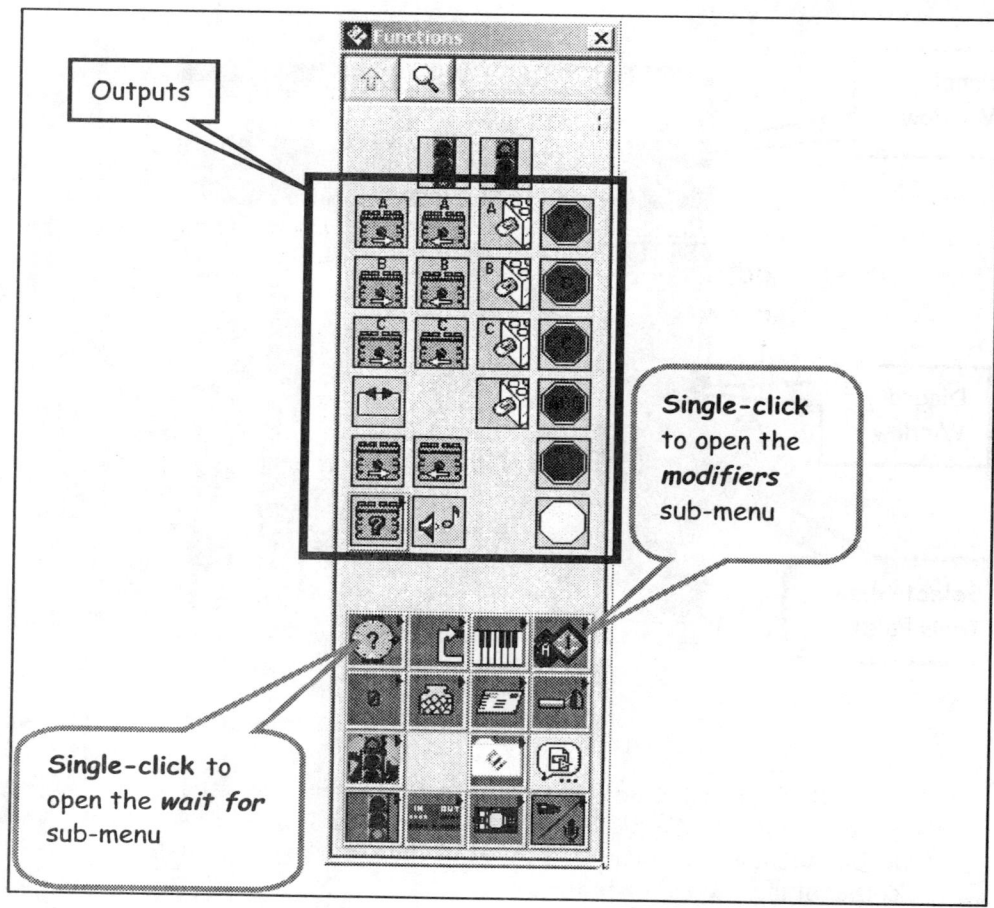

Figure 2.24
Inventor 4 Functions Palette.

2.6.2 The Tools Palette

The Tools Palette is not shown by default, but is very useful for beginners. To show the Tools Palette, select "show Tools Palette" from the Windows Menu (see Figure 2.22).

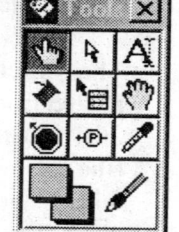

Figure 2.25
Tools Palette.

> **TIP**
> Pressing the SPACE BAR toggles between the Select and String tools.

- The String Tool is used to "wire" icons together. Get familiar with it, because you will use this tool a lot.

 The Select Tool lets you pick and place or drag icons around. The Select tool can also be used to resize text boxes (see the Text tool below). The Select and String tools are the most commonly used tools.

This is the Change Value Tool. With this, you can change numeric values.

The Text Tool lets you change values, just like the Change Value Tool. However it also lets you add text boxes to your program – which is very useful for **commenting** your program.

> In computer programming lingo, **commenting** is the process of adding text comments to your program so that others can understand what you are doing.

> **TIP**
> Pressing the TAB key switches between the Select, Placement, Text, and String tools.

- The Placement Tool lets you move around the contents of the diagram window. It also can be used to pick and place icons from the **functions palette** to the **diagram window** (just like the Select Tool). You can also use the scroll bars to move the **diagram window** contents around.

- Shortcut Tool. This is the same as right-clicking on an icon. In ROBOLAB, the only really useful shortcut is **replace**. Deleting an icon often causes broken wires (Figure 2.26). The **replace** shortcut allows you to substitute one icon for another, which comes in handy since you won't have to re-wire any connections.

> **TIP**
> CTRL+B will delete all bad wires.

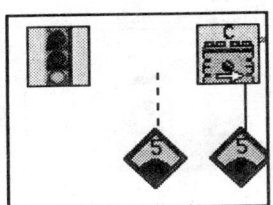

Figure 2.26
Example of a broken wire.

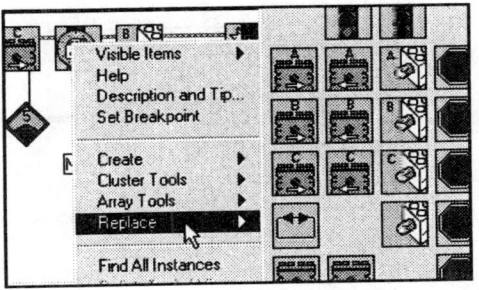

Figure 2.27
The **replace** shortcut.

 **Wiring**

Many of the icons in Inventor mode look familiar from Pilot mode. However, in Pilot mode you never had to deal with wiring the icons together. Since ROBOLAB is a graphical programming language, we need a way of telling the computer in which order to compile our program. We do this by wiring it with the pink string.

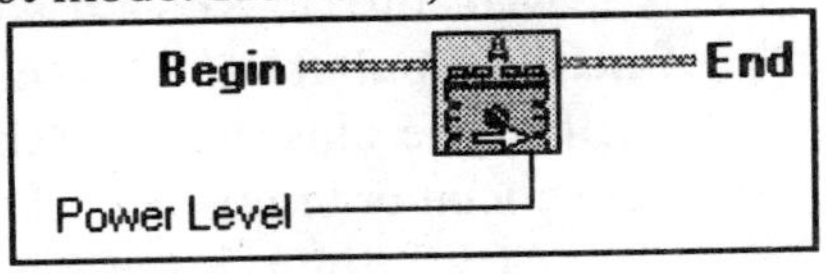

Figure 2.28 shows the general layout of a ROBOLAB function. The **End** terminal of one function is wired to the **Begin** terminal of the next function. **Begin** and **End** terminals are always wired with the pink wire. *Modifiers* will be wired with blue, green or orange wires.

Figure 2.28
Wiring terminals.

Sometimes you need to use a circuitous route for the wire. By clicking anywhere other than on an icon with the String Tool, you can create bends in the wire as shown in Figure 2.29. If you need to stop wiring for any reason, hit the ESC key. Another neat trick is to hit the space bar to flip the wire corner.

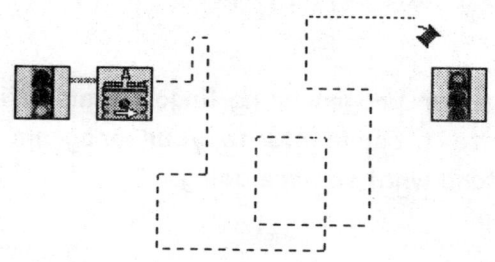

Figure 2.29.
Create bends in a wire by clicking on any blank space.

Figure 2.30 is an example showing how wiring, not the layout of the icons, controls the order of execution. Notice how the three **comments** (text boxes) that were created with the Text Tool help you understand the program.

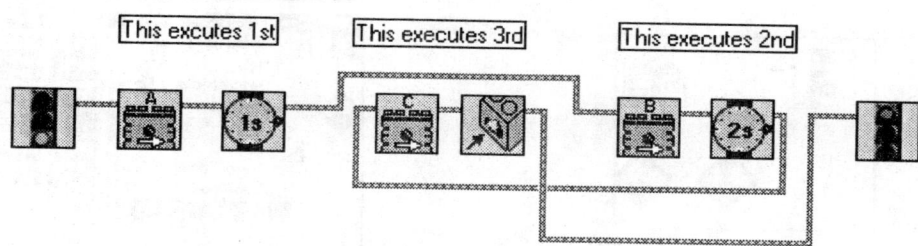

Figure 2.30
Wiring example.

Autowiring

A new feature in ROBOLAB 2.5 is **autowiring**. When you pick a program icon from the **functions palette** and place it onto the **diagram window** it attempts to automatically wire itself to the closest icon. This is usually very helpful, but sometime it wires itself to the wrong icon or terminal.

By hitting space bar once while you are dragging an icon around, the icon will attempt to autowire itself to the nearest icon (note, you must already be dragging the icon around before pressing the space bar). This is useful for autowiring icons that are already in the **diagram window**.

2.6.3 Getting Help

The context help window is the most beneficial feature of ROBOLAB for beginners. You can access Context Help from the **Help** menu as shown in Figure 2.31.

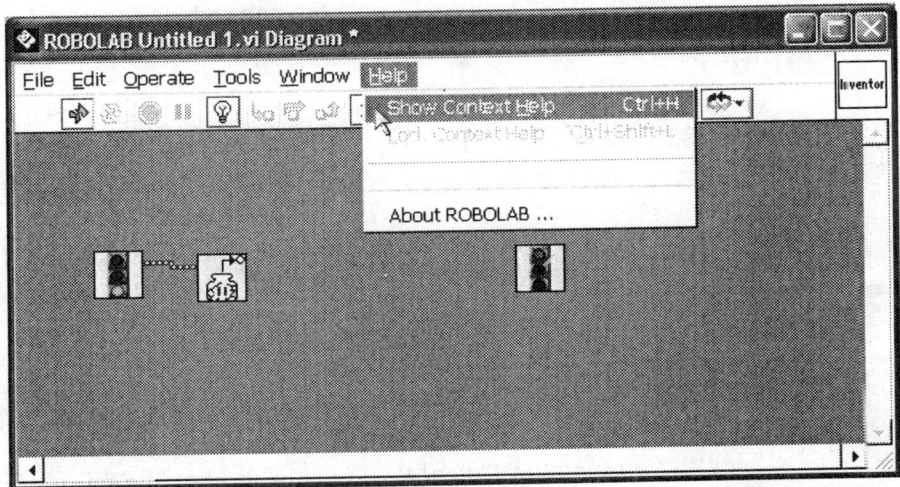

Figure 2.31. The context help is access from the **Help** menu

Context Help provides information on where to wire the functions. All of the different connection points (called terminals) are color coded for easy identification. For example port modifiers are connected to functions using green colored wire. Integer numeric constants use blue wire. Floating point numeric constants use orange wires.

The help window also lists all the **default values** for a function. The default values are the values used when no modifiers are wired to the function.

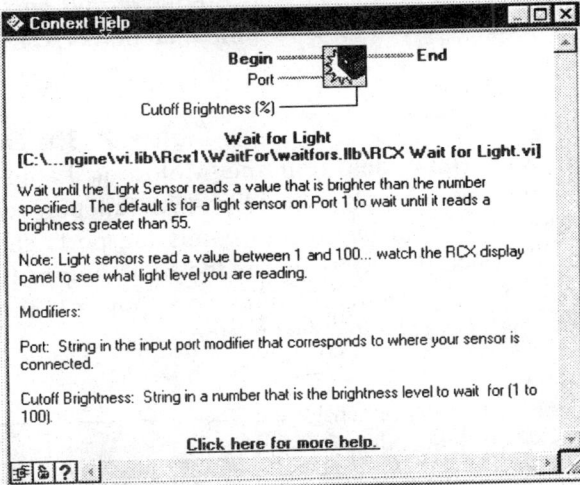

Figure 2.32. The context help window is <u>very</u> useful. When active, placing your cursor over any ROBOLAB icon will lead to detailed help.

2.7 The Basic Inventor Badge

The purpose of this section is to provide you the information necessary to earn the Basic Inventor skill badge. Like Pilot Mode, we are going to jump right to Inventor Level 4.

2.7.1 Outputs

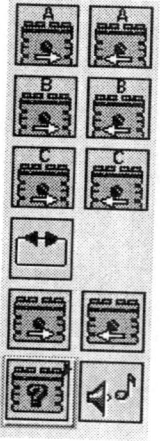

Figure 2.33. Motor outputs.
The six main motor functions are basically identical to those in Pilot mode. As will be discussed in section 2.7.3, the only difference is that a *power level modifier* has to be wired to the motor icon as shown in Figure 2.28. If a *modifier* is not wired to a motor, it will by default run at full power (power level 5).

There are also several new motor functions in Inventor. The first is *flip direction* which reverses the direction of rotation of the specified motor(s). The next two functions are the generic *motor forward* and *motor reverse*. Both the power level and the output port(s) are specified using *modifiers*, as will be discussed in Section 2.7.3.

Using the *play sound* function you can also play one of six sounds on the internal speaker.

Finally, the *advanced output control* sub-menu has its own skill badge and will be covered in Chapter 4.

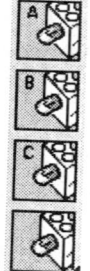

Figure 2.34. Lamp Outputs.
The *Lamp A, B, C* outputs are similar to the Motors in that the *power level* modifier should be wired to them. If no modifier is used, they default to power level 5 (full power) In addition to the *power level* modifier, the generic *Lamp* output also requires that you wire the *output port* modifier. If no modifiers are wired, the default is to turn on all ports (A, B, and C) at full power.

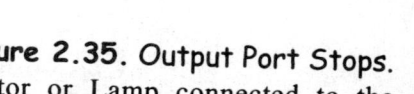

Figure 2.35. Output Port Stops.
Stop A, B, C functions stop the Motor or Lamp connected to the corresponding output port. The *Stop ABC* function stops all three output ports. The generic *Stop* function stops the ports specified by the *port modifiers*. The yellow stop sign is the *float output* function. Unlike the stop functions, the float output simply cuts power to the output port, which will allow motors to slowly coast to stop.

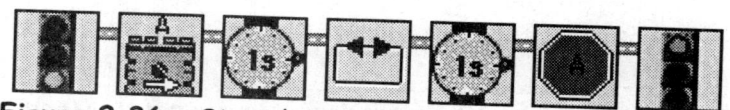

Figure 2.36. Start/stop motors example. Note, the default values are used since no modifiers are wired.

The program in Figure 2.36 turns on Motor A in the forward direction waits for 1 second, flips the direction of Motor A, waits for 1 more second, and stops Motor A before ending.

Note that no modifiers were used (Inventor mode modifiers will be discussed shortly). In ROBOLAB, all functions have **default values** that are used if no modifiers are wired to them (the default values are always listed in the Context Help window as shown in Figure 2.32). In this case, the default power level for Motor A is full power and the default was to flip the direction of all output ports.

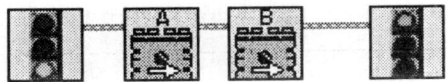

Figure 2.37
The motors won't stop when program ends!

Note the importance of the final *Stop A* in figure 2.36. Without it Motor A would continue to run even though the program ended. For example, in figure 2.37, both Motors A and B will continue running until either the **prgm** button on the RCX is pressed or the RCX is turned off. This program will also execute in essentially no time (a few milliseconds).

Figure 2.38. This motor demonstrates the behavior of the *float* function.

The difference between *Stop* and *Float* functions is best exemplified with a couple of little experiments. Figure 2.38 shows a motor with no lead wire attached to it. Try spinning it and notice how easily it turns and continues to spin for little while, coasting to a stop. This is equivalent to the *float* function.

Figure 2.39. Using a LEGO motor as a generator.

Now let's try a second experiment. Figure 2.39 shows two motors connected together with a lead wire. Turn one of the motors and the other will miraculously turn! What's going on? The motor you turn is acting as a generator and the power is being used to active the other motor. In the first experiment (Figure 2.38) the motor also acted as a generator, but there wasn't any *load* (i.e. nothing was using the power generated) so we didn't notice that we generated any power. The orientation of the lead wires will also determine the direction the motor turns; see if you can get the motor to turn the same direction and the opposite direction as the generator.

Figure 2.40. This motor demonstrates the behavior of the *stop* function.

Our last experiment illustrates the concept of an electric brake, which is what happens when you use the *stop* function. In Figure 2.40, both ends of the lead wired are stacked at 90°. Try spinning it and you will notice that it is very hard to turn. What going on? In this experiment we are using the motor as both a generator and a motor *at the same time*. By spinning the motor we generate electricity. The electricity passes though the wire and powers the motor to spin – in the opposite direction. The end result is an electric brake.

When you use the stop function, the RCX cuts the power to the motor AND short circuits the motor to create an electric brake, just as we did in the third experiment (Figure 2.40). When you use the float function, the RCX cuts the power but leaves motor circuit open, just as we did in the first experiment (Figure 2.39).

 ## 2.7.2   The "Wait For" Functions

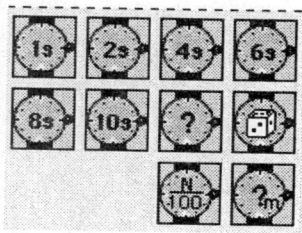

Figure 2.41. Time wait for functions.
The first six *wait for time* functions specify the time the program will wait in seconds before proceeding to the next function. For the watch icon with the question mark, the number of seconds is specified by wiring a *numeric constant* modifier to the function. Similarly for the *wait for random time* function, the *numeric constant* modifier specifies the maximum time to wait. The last two *wait for time* functions allow you to specify the time in hundredths of a second and minutes respectively (again using the *numeric constant* modifier).

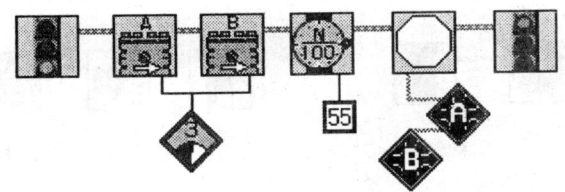

Figure 2.42. Sample time wait for.

The program in Figure 2.42 will turn on both Motors A and B in the forward direction at power level 3, wait for 0.55 seconds, and then float Motors A and B. Because the *float output* function was used, the motors will slowly coast to a stop when the program ends. Note we have taken a little bit of liberty here because we've used several **modifiers**, which will not be fully explained until section 2.7.3 later in this chapter.

Figure 2.43. Touch sensor wait for functions.
The **wait for push** function is nearly the same as in Pilot mode. The two exceptions are that 1) you have to wire the **port** modifier to specify the port and 2) you can also specify the number of pushes to wait for with a **numeric constant modifier**. For example, you may want to wait for 3 pushes before proceeding to the next function, as shown in Figure 2.44. The **wait for release** function requires only the port modifier.

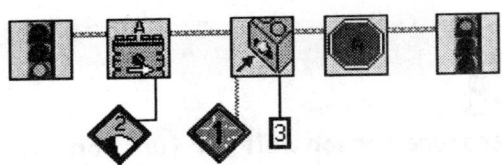

Figure 2.44. Sample wait for multiple push program.

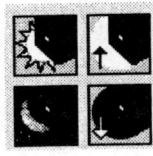

Figure 2.45. Light sensor wait for functions.
The **wait for light** function waits until the light sensor value is higher than the specified value, which is defined by using a **numeric constant** modifier (100 = bright, 0 = dark). Similarly, the **wait for dark** function waits until the light sensor value is less than the specified value.

The **wait for lighter** and **wait for darker** functions wait until the light sensor reading is brighter or darker than the <u>current</u> value. Both the **input port** modifier and the amount of change must be wired to these functions.

It is important to note that the first two functions are based on absolute light sensor values whereas the later two functions are based on relative changes in the sensor values. This is important for accommodating varying ambient light conditions. For example, at home you may be dealing with one ambient light level and at school another. If you used the wait for light or dark functions you will probably get very different behaviors at home and school because of the difference in ambient light levels.

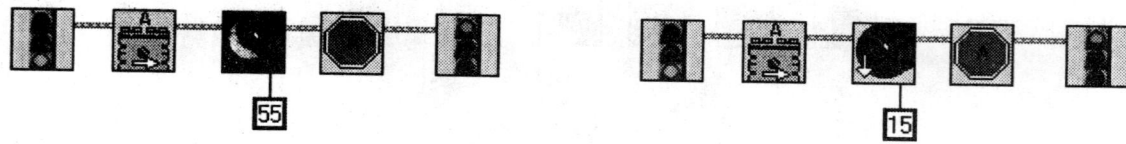

Figure 2.46. Wait for dark (absolute) and wait for darker (relative change) functions.

The program on the left of Figure 2.46 turns on Motor A in the forward direction at full power (the default value), waits for the light sensor on input port 1 (the default value) to drop below 55 before turning off Motor A and ending. The program on the right is nearly the same except that it waits for the light reading to drop by 15 before turning off Motor A and ending. Again, in both programs we've taken advantage of the default values built into the ROBOLAB functions to avoid having to wire *modifiers* to them.

One interesting feature of the *wait for light* functions is that the LCD will switch from showing the clock to viewing the light sensor reading. This also true for both the *wait for temperature* and *wait for rotation* functions (see below). This can be very helpful when trying to debug your programs.

The next two sets of wait for functions require the temperature and rotation sensors, which is not included in the standard ROBOLAB Team Challenge set but can be purchased separately.

Figure 2.47. Temperature sensor wait for functions.

The *wait for decreasing* and *wait for increasing temperature* functions cause the program to pause until the temperature is below or above the specified temperature respectively. Either Celsius or Fahrenheit temperature scales can be used. All four functions require both the *input port* and *numeric constant* modifiers to be wired to them to specify the sensor input port and temperature. A sample program using a pair of temperature wait for functions is shown at the end of this chapter in Figure 2.61.

Figure 2.48. Rotation sensor wait for functions.

The *wait for rotation* function waits for the rotation sensor to exceed the number of rotation units specified. There are 16 rotation units per revolution. Similarly, the *wait for angle* function waits for the angle to exceed the specified value. Since there are 16 divisions per revolution, the angular resolution is 22.5 degrees. The last rotation wait for function, *wait for rotation w/o reset*, is similar to the wait for rotation function except that the rotation sensor is not zeroed each time. Both the *input port* and *numeric constant* modifiers are used in conjunction with all rotation sensor wait for functions.

Figure 2.49. Sample wait for rotation function.

The program in Figure 2.49 turns on Lamp A and then waits for the rotation sensor on input port 1 to exceed 16 rotation units (1 full rotation) before turning off the Lamp A and ending. Note no modifiers were wired to the function so the default values, input port 1 and 16 rotation units, were used.

Figure 2.50. Other wait for functions.
The rest of the wait for functions and sub-menus will be covered in Chapters 3 and 4.

2.7.3 Modifiers

We've used modifiers several times in Pilot mode and in the examples above, but we still need to formally introduce them and explain their use in Inventor mode. Modifiers are used to specify input ports, output ports, power levels, and numeric constants. Modifiers are picked and placed from the **functions palette** to the **diagram window** just like other ROBOLAB icons.

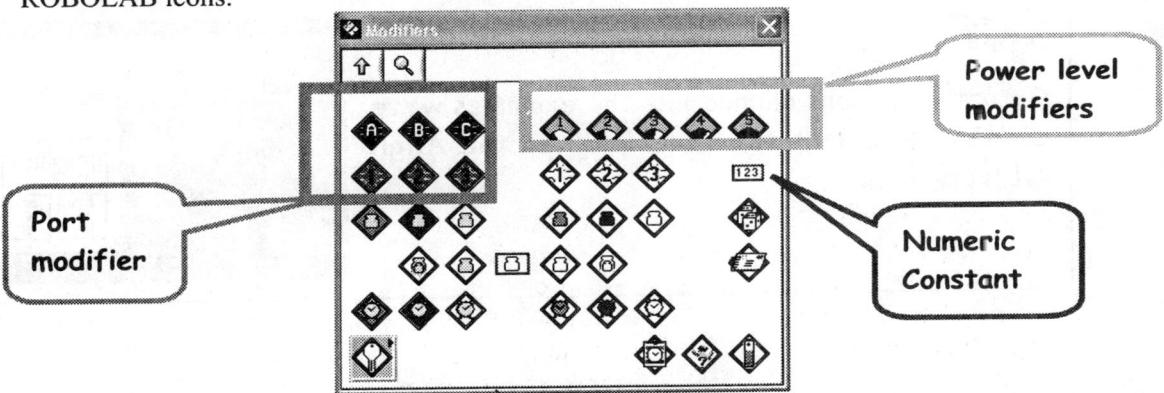

Figure 2.51. The modifiers sub-palette. For now, we'll only be using a few of the icons.

You won't find an orange numeric constant on the modifiers sub-palette; the integer numeric constant will change from blue to orange automatically when a floating point number is entered.

Figure 2.52. Examples of bad wiring.
Left: the modifiers are wired to the wrong terminals, which leads to broken wires.
Right: an input port modifier (1, 2, 3) is wired where an output port (A, B, C) should be, which does not lead to a broken wire (less obvious mistake).

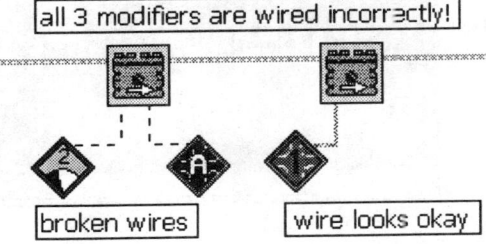

Figure 2.53. Broken Run Arrow.
If you have wiring errors, the Run arrow (upper left) will appear broken. Clicking on the broken arrow will open a pop-up window listing all the errors.

Figure 2.54. Integers and floating point numeric constant modifiers.
Both *wait for* functions will pause for the same length of time, 0.31 seconds. The first *wait for* uses an integer numeric constant modifier (blue) while the second uses a floating point numeric constant modifier (orange).

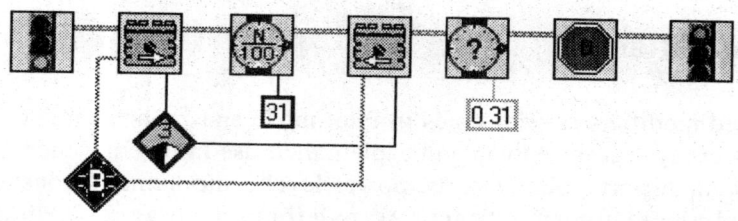

 To get your icons aligned like the examples we've shown, use select the icons and use the **Align Objects** menu.

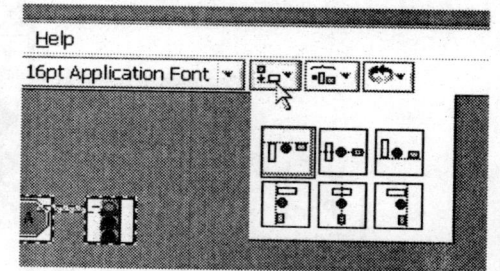

2.7.4 Getting to the Source

If you're interested, you can get to the source code for any of the ROBOLAB functions by double clicking on any function, as shown in Figure 2.55.

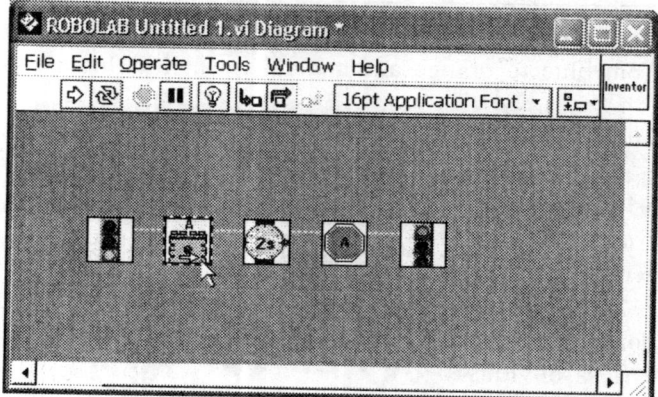

Figure 2.55. Double click on any icon to access the source code. In this example, we're opening the *Motor A forward* function.

Double clicking an icon will open the **panel** for that function. Figure 2.56 shows the **panel** for the *Motor A Forward* function. To see the **diagram** window (which is where we normally do all the wiring), select **Show Diagram** from the **Window** menu.

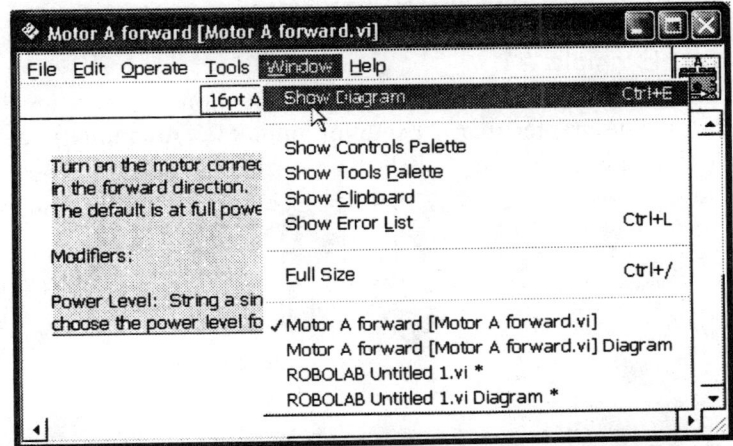

Figure 2.56. On the Window menu, select Show Diagram.

Once the diagram window is open, you can see how the function is put together and where the modifiers come into play. What you are seeing is the LabVIEW G-Code, which is covered in Chapter 8.

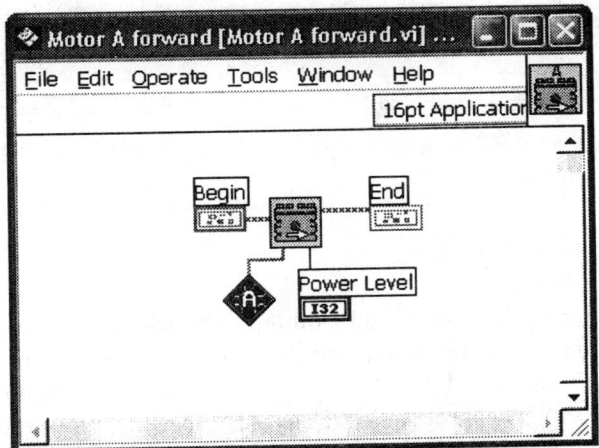

Figure 2.57. The *Motor A Forward* diagram.

2.7.5 Sample Programs

The following 4 programs will help you gain an understanding of basic Inventor programming. If you can read and understand what they do, you should have no problem earning the Basic Inventor skill badge.

Figure 2.58. Example #1.
This program turns on Motor A in the forward direction at power level 3 until the rotation sensor reads a value greater than 24 rotation units (1.5 revolutions) at which time it turns off Motor A and ends.

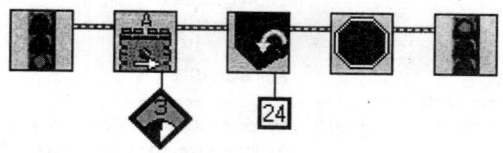

Figure 2.59. Example #2.

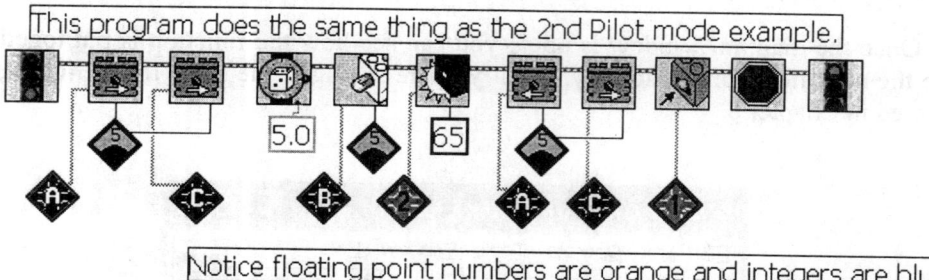

This program is the same as the Pilot program shown in Figure 2.11 done in Inventor mode.

Figure 2.60. Example #3.
This program turns on Motors A and B in the forward direction at full power until the touch sensor on input port 1 is pressed. Then it reverses both motors for 1 second, stops both motors and then turns on Motor A in the forward direction at power level 3 for a random amount of time (up to 2 seconds). Motor A is then stopped and the program ends. It is always good programming etiquette to stop the motors before ending the program.

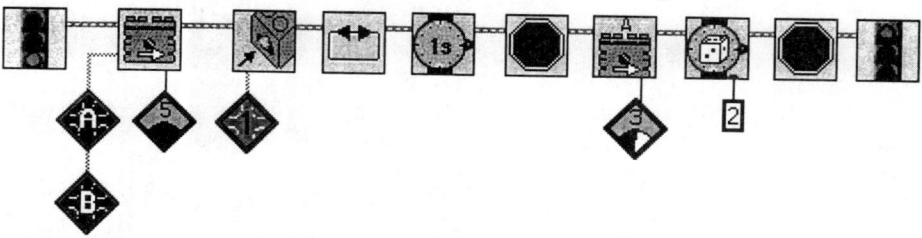

Figure 2.61. Example #4: simple feedback control.

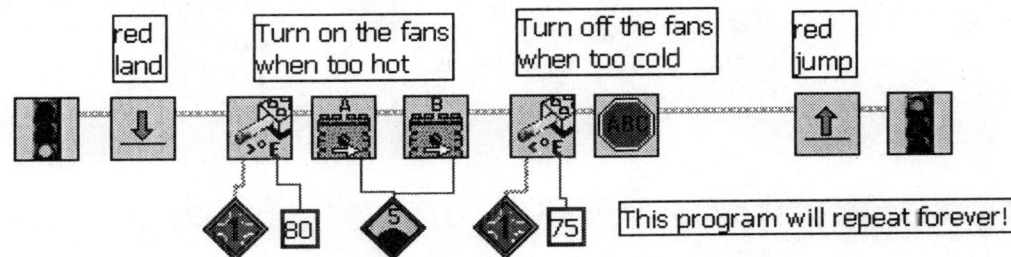

This program uses a *jump/land* pair of functions that you haven't seen yet that form a loop. When the program reaches the *jump*, it "jumps" backwards to the *land*, thus causing everything between the *jump* and *land* to repeat over and over (indefinitely in this case). This program could be used to control a set of fans which maintains the temperature between 75 and 80 degrees Fahrenheit.

2.7.6 What's Next?

Thus far, we are still using Inventor to perform sequential programming. We have not made use of the ability to implement **control structures** yet. This will be the focus of Chapter 3.

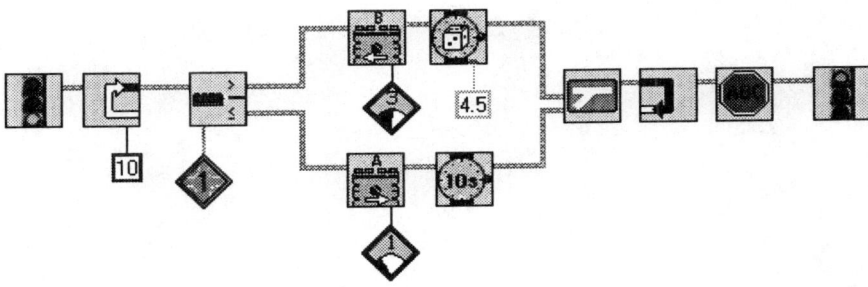

Figure 2.62. Inventor control structures.

CHAPTER 3

WHITE LEVEL

Skill badges available in this Chapter

- Structures
  - Jumps
  - Loops
  - Forks
- Containers
  - Math
  - Timer
  - Sensor
- Tasks
  - Split
  - Start/stop
- Basic Investigator
  - Program
  - Upload
  - View
  - Compute

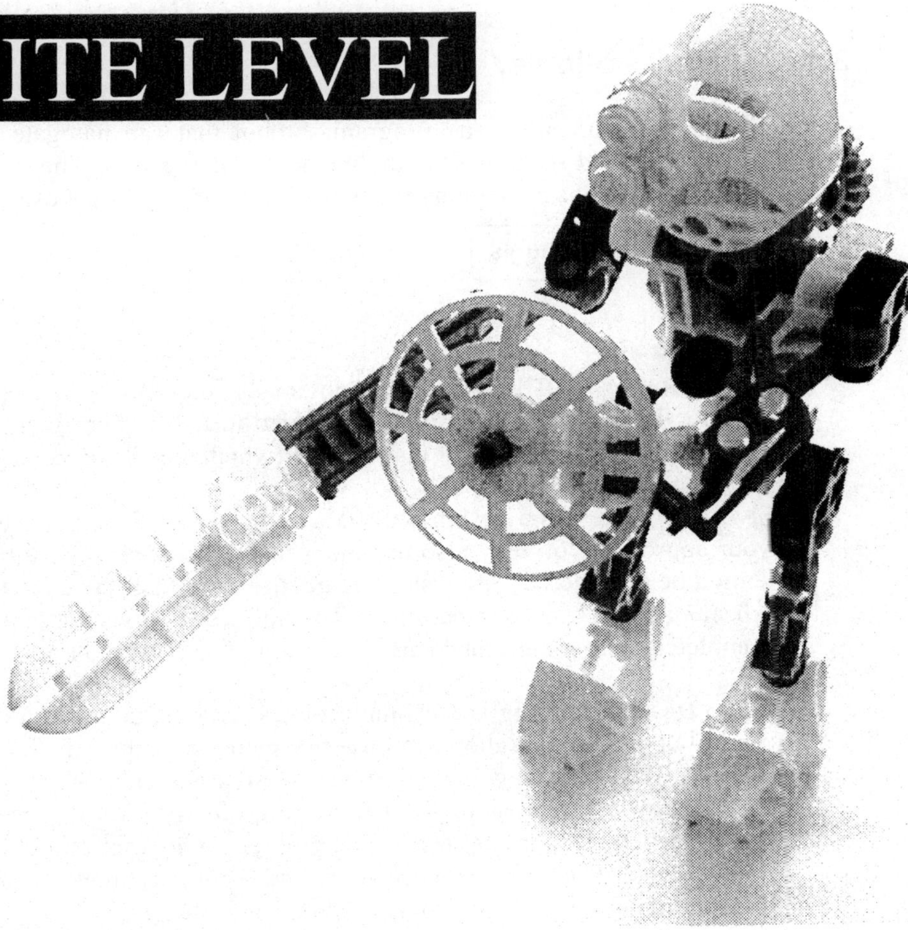

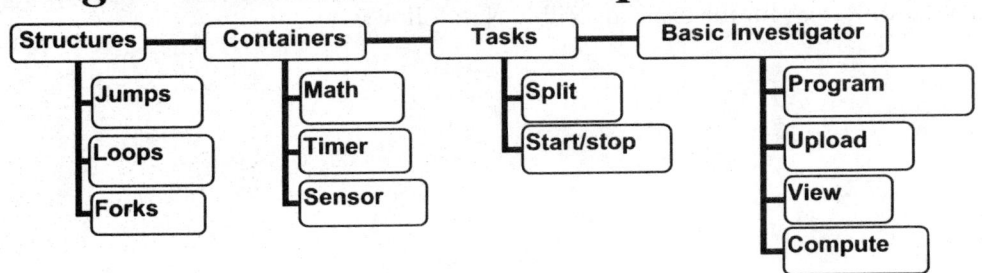

3.1 White Challenges

All of the White Level Design Challenges presume that you have already earned all the Green Level skill Badges.

3.1.1 Wall Follower

Challenge: Design, build, and program a robot that can navigate a maze. You are not allowed to use dead reckoning as the method of navigation. The instructor may give you additional instructions concerning the type of program you must use.

Skill Badges: | Structures |

Procedures:

Experimental Setup: The Challenge obviously requires a maze. A typical maze is constructed out of 2x1 wood boards nailed to a 4x4 piece of plywood. The course is more like a twisting corridor than a maze (which may have forks and dead-ends).

Robot Design: The basic design will be a vehicle with 2 or more wheels. Depending on your approach, you may also use more than 2 sensors. Both turning radius and speed should be considered. The faster you go, the more likely you are to get lost, so consider wheel size and gear ratio carefully. The only restrictions are that your vehicle must be completely autonomous and carry everything (i.e. leave nothing behind).

Program: **Dead reckoning** is the simplest but least reliable method of navigating a known maze, but you are not allowed to use this method for this Challenge. Dead reckoning is accomplished by navigating based on time. For example, when you know the layout of the maze, you can program your robot to go forward for 2 seconds, turn left, forward for 1 second, turn right, etc. The problem with this method is that mistakes are compounded. If a wheel slips early on, the robot may think it's going straight, when it actually may be veering off to one side.

For this Challenge you will use **wall following.** You can navigate some mazes by following either the left or right wall the entire time. Wall following can be accomplished using 2 sensors (either the light and a touch sensor or 2 touch sensors), one on the front of the robot and one on the side. If we designate the front sensor as `front` and the side sensor as `left` (assuming we want to hug the left wall), a flowchart for the basic algorithm is as follows:

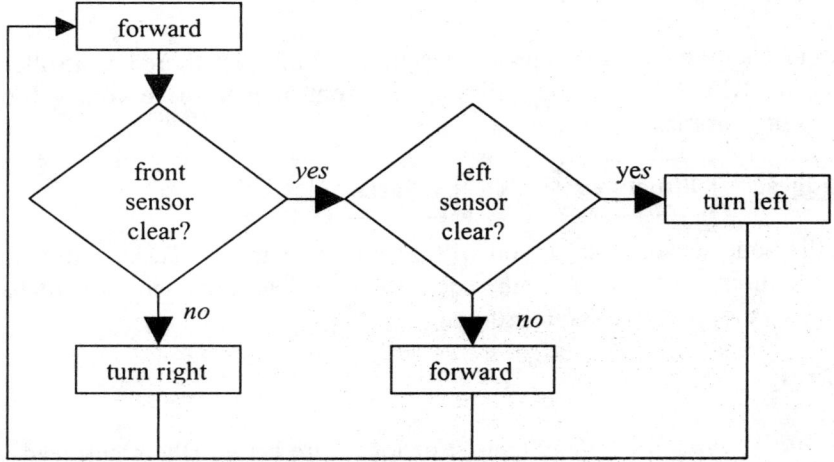

This approach uses a nested *If* function. *If* functions are *forks* in ROBOLAB. The repeat function is a *loop* or *jump* in ROBOLAB. If you use this approach, you will earn the Structures skill badge.

<u>Grading</u>:

Your grade will be based 75% on performance and 25% on creativity and aesthetics.

Performance	Creativity & Aesthetics
A+: You make it through the fastest	A+: Best of show
A: You make it to the end	A: Outstanding
B: You make it more than ½ way	B: Good
C: You seem to headed in the right direction	C: Okay
D: You wander aimlessly	D: Nothing special
F: You don't show up to class	F: Divert your eyes!

3.1.2 Remote Controller

Challenge: Using at most 2 sensors, design and build a tethered controller for remotely operating an RCX vehicle. You will use your controller to drive your vehicle through an obstacle course or maze as fast as you can.

Skill Badges: | Structures | AND | Tasks |

If more than one sensor is used, you will most likely earn the Tasks skill badge in addition to the Structures skill badge. Your instructor will determine which skill badges you get based on an examination of your program.

Procedures:

Experimental Setup: The tethers consist of long wire bricks (the black lead wires). LEGO sells 10 foot long versions, or several shorter ones can be connected in series. An obstacle course or maze is also required for students to demonstrate controlled driving of the vehicle.

Robot Design: Nearly any vehicle will suffice for this Challenge. You should consider designing a controller that can control both speed and direction. The only constraints are that the RCX must be on your vehicle (i.e. no holding it in your hand) and you may use at most 2 sensors on the remote control. Non-LEGO parts, such as paper and tape, may be allowed at the instructor's discretion.

Program: Your program will consist of **sensor forks**. You will also most likely use **loops** or **jumps** in your program to continually monitor the sensor(s). If you use more than one sensor and wish to monitor both of them simultaneously, you will also have to use **tasks**.

Grading:

Your grade will be based 80% on performance and 20% on creativity and aesthetics.

Performance	Creativity & Aesthetics
A+: You make it through the fastest	A+: Best of show
A: You make it to the end	A: Outstanding
B: You make it more than ½ way	B: Good
C: The robot is under your control at some point in time	C: Okay
	D: Nothing special
D: You wander aimlessly	F: Divert your eyes!
F: You don't show up to class	

3.1.3 LEGO Brick Recycler

Challenge: Design, build and program a machine for sorting LEGO bricks based on color.

Skill Badges: | Structures |

Procedures:

Experimental Setup: The only items required for this Challenge are a few LEGO 2x4 bricks: 4 black, 4 white, and 2 blue. The blue bricks are the "challenge" bricks – the hard ones to sort.

Robot Design: The operator (the instructor) will place the bricks into your machine in a random order. Your machine should move the bricks in front the light sensor and then sort them into one of three bins based on color. More creativity points will be given to designs that do not require the operator to load or orient the bricks in a special way.

You will use the light sensor and one or more motors for this Challenge. You may also consider using a touch sensor.

Program: Forks and loops will be used to complete this Challenge. In order to sort all three colors, you will need to use nested forks.

Hints: The LEGO light sensor is very sensitive to ambient light levels. You may want to consider shielding the light sensor from the room lights as in Figures 2.12 and 2.13 or even under some kind of shield. Also, taking multiple (repeated) light readings will increase accuracy.

The light sensor is also very dependent on the distance between the sensor and the brick (it is actually a very good proximity sensor). For example, a black brick very close to the sensor will look like a white brick that is far away.

Grading:
Your grade will be based 75% on performance and 25% on creativity and aesthetics.

Performance	Creativity & Aesthetics
A+: Sort the most bricks correctly the fastest	A+: Best of show
A: Sort 9 bricks correctly	A: Outstanding
B: Sort 6 bricks correctly	B: Good
C: Sort 4 bricks correctly	C: Okay
D: Move the bricks around	D: Nothing special
F: You don't show up to class	F: Divert your eyes!

3.1.4 Stay Inside the Box

Challenge: Construct, completely from your LEGO kit, an autonomous robot capable of staying within a square denoted by four lines taped on the floor. While inside the box, your robot is to perform the most interesting task possible.

Skill Badges: | Structures |

You may earn additional badges, depending on your program. Your instructor will decide.

Procedures:

Experimental Setup: All that is required is four lines of a contrasting color to the floor. Black electrical tape on a white tile floor works well. The instructor will specify the location of Challenge so that you know what color the floor is.

Robot Design: In order to perform useful tasks, a robot must be able to properly navigate its way through its surroundings. In order to move through its environment, a robot may be called upon to avoid obstacles, detect changes in light or heat, or find its way through a maze of hallways. The American Nuclear Society held a competition several years ago specifically designed to test a robot's navigation and object detection abilities. They mapped out a grid of white tape on a black floor to aid the robot in its navigation. Those who could take advantage of these contrasting lines were able to move about the arena more easily.

There are many ways to stay inside the box, the most obvious of which is to mount a LEGO light sensor on your robot. Where and how you mount your sensor(s) is entirely up to you. There is no limit on the weight or size of the robot, although originality and creativity will inevitably win you praise (but not necessarily points) from the judges. The only restriction is that your robot may not stop moving for more than two seconds. In other words, a robot that never moves will perform the task (i.e. it will never leave the box), but will earn you a D grade.

You are also responsible for **programming** your RCX to perform this task. Some suggestions for interesting tasks:
- **Run and Stop** - Drive towards one line on the floor, stop once you reach it, turn around and run again until you reach another line. (B)
- **Line-Following** - Start your robot in any orientation, find one of the lines on the floor and follow it all the way around the perimeter. Simple. (B+)
- **Obstacle avoidance** - Start in any orientation you wish, and perform the run and stop algorithm while avoiding some light-colored objects placed within the box. (A)
- **Your Imagination** - Impress us with something more complex? (A+)

Program: It's totally up to you!

Grading:

Your grade will be based 100% creativity and aesthetics. Your grade will depend entirely on how impressed the instructors are! If your robot doesn't complete the any task but stays in the box you will get a D for the Challenge.

3.1.5 Fetch the Light

Challenge: A flashlight will be placed somewhere on the floor pointing in the direction of your robot. Find this light and get as close as possible to it and stop before running into it. Simple.

Skill Badges: | Structures |

Procedures:

Experimental Setup: Bring your own flashlight. This is critical, since you will be testing with your own flashlight at home. We will have one, but the chances of your robot performing equally well with two different light sources are slim.

Robot Design: A national Fire-Fighting robot competition is held at Trinity College each year in which the entries are required to find and extinguish a fire somewhere in a house. The fire is actually a candle and the house is little more than a simplified maze so the task may seem easy. The trick is that, once the robot has identified the room containing the fire, it must get within 12 inches before trying to extinguish it. Generally, the closer you can get to the candle without touching it, the better. If a robot is running Lego motors and fans, for instance, the closer to the candle it can get, the better chance it has of putting it out.

We will turn the lights off in the room, so that the point of greatest light should be the flashlight, but there is no guarantee that the room will be completely free of natural light. This competition will not be run as a group, so leave the claws of death at home. Everybody will have an unobstructed path to the light source. If you can solve this problem with obstacles, you get and extra half-grade (A+ max).

Program: Simple code always triumphs. Think of some motions your robot could do to solve a simpler problem (say, aim at the brightest spot in the room) and loop through those motions until you have reached a maximum value.

Grading:
Your grade will be based 75% on performance and 25% on creativity and aesthetics.

Performance	Creativity & Aesthetics
A+: Stops less than six inches away	A+: Best of show
A: Your robot stops within one foot of the flashlight	A: Outstanding
	B: Good
B: Your robot stops within 3 feet of the flashlight	C: Okay
	D: Nothing special
C: Your robot moves towards the light	F: Divert your eyes!
D: Your robot is light-shy	
F: You don't show up to class	

3.1.6 Robot Zoo

Challenge: This is not a competition in the normal sense. Rather, we want you to build your interpretation of an animal with LEGO bricks. If you visited a zoo full of LEGO animals, what would they look like? How would the move? Be creative, your imagination is the only limitation this week.

Skill Badges: Structures

Procedures:

Experimental Setup: All you need is a large enough area to let all your animals roam free.

Robot Design: Some things to think about:
- Most animals don't move with wheels. If yours has legs in real life, it should probably have them in LEGO life.
- If your animal moves with legs, how does its gait look?
- Does this animal have knees? Elbows? Shoulders?

Just like in nature, Darwinism applies. We will have all of your creations have a go at each other in a full-on mass competition where survival of the fittest is the only rule

Program: You will most likely use structures in this Challenge, if for no other reason to make the program loop over and over. You may earn other skill badges based on the complexity of your program (you instructor will decide).

Grading:
Your grade will be based 20% on performance and 80% on creativity and aesthetics.

Performance	Creativity & Aesthetics
A+: Top of the food chain!	A+: Best of show
A: You play fair & have fun	A: Looks like the real thing
B: You play fair	B: Looks like an animal
C: Your animal moves and makes noise	C: Something moves around
D: You animal moves	D: Looks like it was once alive
F: You don't show up to class	F: Put it out of its misery!

3.1.7 Edge Detector (a.k.a. Barcode Reader)

Challenge: Design and build an edge detector than can sense the transition between black and white. Your device should play one sound for a black-to-white transition and another sound for a white-to-black transition.

Skill Badges: Structures AND Containers AND Music

Note: For information on the Music skill badge, see Chapter 4.

Procedures:

Experimental Setup: A strip of white paper with black stripes of varying thickness and spacing is required. The instructor will provide the dimensions of the paper strips to be used. A variation (to add difficulty) is to include several shades of gray, which also must be detected.

Robot Design: Your device will have to either pull the paper strips in, move the light sensor over a stationary strip of paper, or a combination of both. The only restriction is that you are not allowed to manually move the paper or light sensor (i.e. the device must be fully automated).

Program: The basic algorithm is to read in the current light sensor value and compare it to the last reading. If there is no difference, then there has not been a transition between colors. If it is different, then a transition has occurred and we should play a sound. You will have to determine what constitutes "no difference" between readings.

Hints: Shading the light sensor from the ambient light condition will result in the most consistent light reading.

The light sensor is also very dependent on the distance between the sensor and the brick (it is actually a very good proximity sensor). For example, a black stripe very close to the sensor will look like a white stripe that is far away.

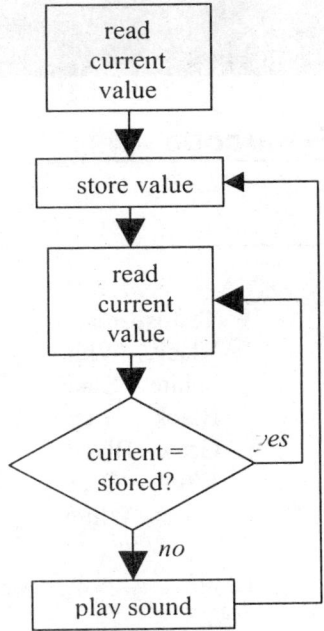

Grading:

Your grade will be based 80% on performance and 20% on creativity and aesthetics (pay attention to the sounds your device makes). The instructor will provide 4 different strips of paper, each with different complexity of transitions. The "A" strip will the most complex, with many transitions. The "D" strip will be the least complex (e.g. half white, half black).

Performance	Creativity & Aesthetics
Your grade will be based on the most complex of the 4 strips you can successfully read. If you do not show up to class, you will get an "F".	A+: Best of show A: Outstanding B: Good C: Okay D: Nothing special F: Divert your eyes!

Representative barcodes

A- barcode

B- barcode

C- barcode

D- barcode

F- barcode

Transition	Note
Black – White	A-note
White – Black	B-note
Black – Grey	C-note
Grey – Black	D-note
White – Grey	E-note
Grey – White	F-note

3.1.8 Round and Round (a.k.a. shaft encoder)

Challenge: Build a shaft encoder to measure the distance traveled by LEGO car. Display the total distance traveled on the LCD display.

Skill Badges: | Structures | AND | Containers | AND | Tasks |

Procedures:

Experimental Setup: This Challenge requires a tape measure and a flat surface.

Robot Design: Shaft encoders are commonly used to measure the rotation and velocity of an axle. The odometer in your car measures the total number of rotations of the wheels and the speedometer measures the angular velocity of the wheels. Encoders can be built from touch sensors that count the number of clicks or a light sensor that counts the number of light/dark transitions. In this Challenge you'll use the light sensor.

Build a LEGO vehicle and either make the an encoder like the one shown in the photo above or attach one of the black and white disks below to a wheel, axle, or gear (or you can create your own black and white disks using some paper and a black pen). Your car should be programmed to move a random amount of time, between 5-10 seconds, before stopping (your car must travel at least 3 feet in 5 seconds). Measure or calculate the distance traveled per black-white transition and use a little container math to convert the number of transitions into the distance traveled.

Program: The program will use at least 2 containers and 2 tasks. One task will turn on the motor for a random time (5-10 seconds) and then stop the car. The other task will be used to measure the distance traveled and display it to the LCD display. You will have to determine what light level corresponds to black and white for your light sensor. A basic encoder flowchart algorithm is shown on the next page. In the flowchart, we've used Container1 to keep track of the last light sensor reading (1 = white, 0 = black) and Container2 to keep a running total of the number of transitions (e.g. a counter).

Hints: The LEGO light sensor is very sensitive to ambient light levels. You may want to consider shielding the light sensor from the room lights as in Figures 2.12 and 2.13 or even under some kind of shield.

Also think about how the speed of your vehicle might affect the accuracy of your measurements.

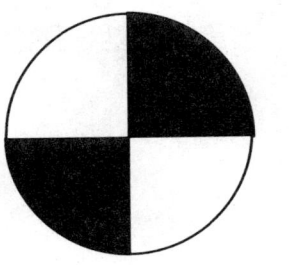 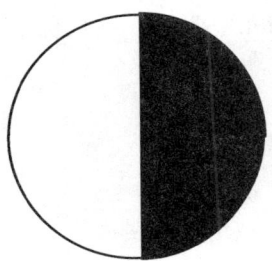

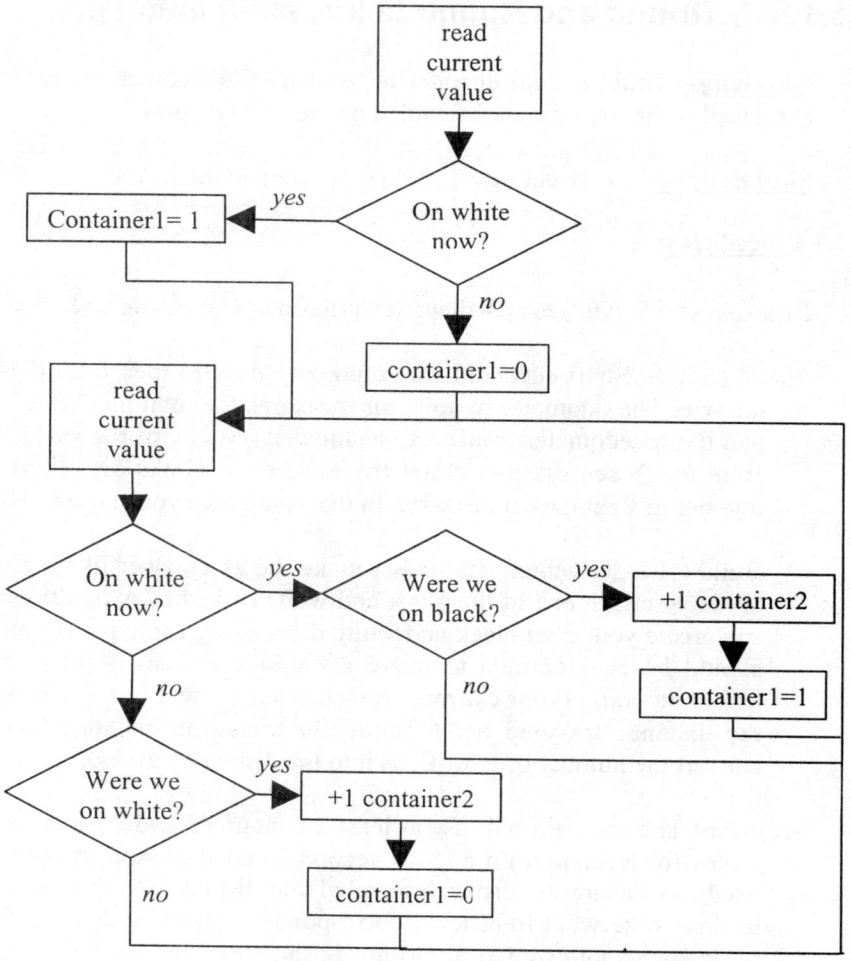

Grading:
Your grade will be based 75% on performance and 25% on creativity and aesthetics.

Performance	Creativity & Aesthetics
A: Your accurate to within 6 inches	A+: Best of show
B: Your accurate to within 12 inches	A: Outstanding
C: Your accurate to within 2 feet	B: Good
D: Your car runs, stops, and displays something to the LCD panel.	C: Okay
	D: Nothing special
F: You don't show up to class	F: Divert your eyes!

3.1.9 Two steps forward, one step back

Challenge: Build a non-motorized car. The instructor will push your car forwards and backwards. Your car should make a noise when the net forward distance traveled exceeds 4 feet.

Skill Badges: Structures AND Containers

Procedures:

Experimental Setup: A flat surface and a tape measure.

Robot Design: Shaft encoders are used in many devices (see Challenge 3.1.8). The problem with shaft encoders is that they cannot differentiate forwards from backwards rotation. However, it is possible to make an encoder that can tell which direction the shaft is turning as well as the position by using 3 colors instead of 2.

Build a LEGO vehicle and attach one of the shaded disks below to a wheel, axle, or gear (or you can create your own disks using an Excel pie chart). Measure or calculate the distance traveled per black-white transition and use a little container math to convert the number of transitions into the distance traveled.

Program: The program will use at least 2 containers. One to keep track of the forward distance traveled and another to keep track of the last light sensor reading.

When the <u>net</u> forward distance traveled exceeds 4 feet, your car should beep (or make some other creative noise). For example, 2 feet forward and 1 foot backward is 1 net foot forward.

Hints: See Challenge 3.1.8 for tips building a shaft encoder. You may find it useful to display the current distance traveled to the LCD.

Grading:
Your grade will be based 75% on performance and 25% on creativity and aesthetics.

Performance	Creativity & Aesthetics
A+: Undefeated champion	A+: Best of show
A: You win more than once	A: Outstanding
B: You put up a good fight	B: Good
C: Motors turn on when touch sensor is pressed	C: Okay
D: You have something to connect the string to	D: Nothing special
F: You don't show up to class	F: Divert your eyes!

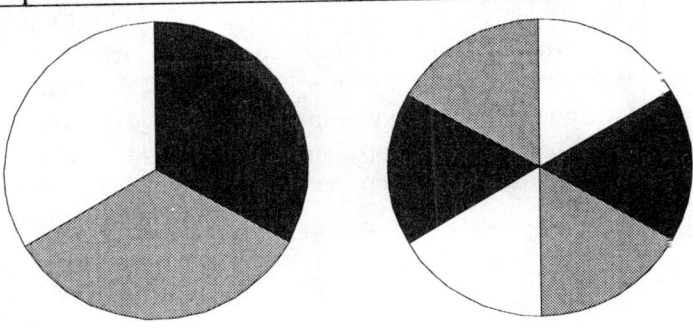

3.1.10 There and back again

Challenge: Build a LEGO car that will drive forward until it bumps into something and then backs up and returns to the start line. The objective is to get as close to the start line as possible. The trick is you are not allowed to use a touch sensor! You are also forbidden from using the light sensor to detect the start line.

Skill Badges: [Structures] AND [Containers] AND [Tasks]

Since there are many ways to complete this Challenge, your instructor will decide which skill badges you earn after examining your program. The most common badges earned are Structures, Containers, and Tasks.

Procedures:

Experimental Setup: This Challenge only requires a flat surface and something to bump into (the instructor's foot works well).

Robot Design: Shaft encoders are used in many devices (see Challenge 3.1.8). By monitoring the rotation of a non-drive wheel, you can use an encoder to detect that you've bumped into something. If you're supplying power to the drive wheels and the non-drive wheel isn't turning, there's a good chance that you've bumped into something and aren't actually moving!

Build a LEGO vehicle and attach one of the shaded disks below to a non-drive wheel. Your car should be programmed to drive forward until the drive wheel(s) start to slip. You should then stop, reverse the motor direction and drive backwards, stopping as close to the start line as possible. Thus, you also need to use the encoder on the non-drive wheel to monitor how far forward you traveled before stopping.

Program: The program will use at least 2 containers and one of the timers. You will have to monitor the time between black/white transitions and stop the motors if too much time elapses between transitions. You will also have to count the total number of transitions on the way out and then use the total on your journey back to the start line.

Hints: See Challenge 3.1.8 for tips building a shaft encoder.

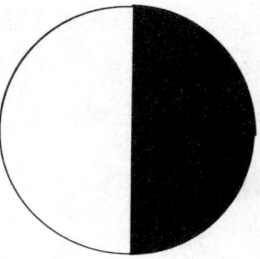

Grading:

Your grade will be based 75% on performance and 25% on creativity and aesthetics.

Performance	Creativity & Aesthetics
A: Your accurate to within 6 inches	A+: Best of show
B: Your accurate to within 12 inches	A: Outstanding
C: Your accurate to within 2 feet	B: Good
D: Your car runs, stops, and backs up	C: Okay
F: You don't show up to class	D: Nothing special
	F: Divert your eyes!

3.1.11 Swinging with Gravity.

Challenge: Measure the Earth's Gravitational acceleration using Inventor.

Skill Badges: | Basic Investigator |

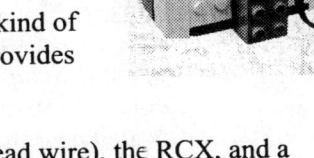

Procedures:

Experimental Setup: You need to swing the pendulum over some kind of light/dark transition. A piece of white paper on a dark table provides an excellent transition (resulting data is a square wave).

Robot Design: Build the LEGO pendulum using some string (or lead wire), the RCX, and a light sensor. An example is shown above. Use a length of string that results in a period of oscillation of about 1 second.

Program: Using Investigator Program Level 1, 2 or 3, write a program to collect light sensor data every 0.05 seconds for at least 10 seconds.

Data Analysis: Upload the data to the computer. View the data and determine the gravitational acceleration on Earth. The period of oscillation of a pendulum is given by the equation:

$$T = 2\pi \sqrt{\frac{L}{g}}$$

where T is the period of oscillation (seconds), L is the length of the pendulum (meters), and g is the Earth's gravitational acceleration (m/s^2).

Thus, if you could measure both L and T of a pendulum, you could calculate the Earth's gravitational acceleration. In this exercise we will build a pendulum of known length (L) and measure the period of oscillation with the light sensor to get T.

Hints: Remember that the formula is only valid for small angles (less than 10 degrees). The length, L, is measured from the point of rotation to the center of gravity of the pendulum (the RCX in this case).

Grading:

Your grade will be based 90% on performance and 10% on creativity and aesthetics.

Performance	Creativity & Aesthetics
A: You get something close to 9.81 m/s^2	A: Outstanding
B: Graph looks normal	B: Good
C: You get a graph and estimate	C: Okay
D: You have data	D: Nothing special
F: You don't show up to class	F: Divert your eyes!

Submit both your graph and your calculations showing how you estimated *g*.

3.1.12 Tomb Raider

Challenge: Determine the path a burglar took around the room using multiple RCX's outfitted with light sensors. This is multi-team project, typically conducted with at least 6 teams.

Skill Badges: Basic Investigator

Procedures:

Experimental Setup: A relatively dark room, a flashlight, and a PC with ROBOLAB are required for this Challenge. The PC may be located in another room, but should be located near the "tomb" to save time.

Robot Design: This is a multi-team project. The goal of this Challenge is to determine, as a group, what path the burglar took around the room by collecting and analyzing light sensor data. The instructor will play the part of the burglar by walking around the room with a flashlight and "tagging" each RCX he/she passes.

Each team should program their RCX to record light sensor readings for 4 minutes. As a group, you will determine how often to take readings. Each team should position their RCX somewhere in the room. Make a map of the room and mark the position of all the RCX's (every RCX has serial number).

Everyone will need to start their programs running at the same time and then leave the room to burglar can go to work. When the burglar is "finished", everyone will return to the room and upload the light readings their RCX collected.

Looking at the graph of light data, determine at what time(s) the burglar passed by your RCX. Mark that time on the map next to your RCX. Find out what time the burglar passed by the other RCX's and record those times as well. Finally, reconstruct the path you think the burglar took around the room.

Program: You must program your RCX to collect light sensor data for 4 minutes. The sampling period will be determined by the collective group.

Hints: Things to consider: how will you synchronize your data? How often should you sample? In the real world, there is usually a cost associated with data collection. Where will you place the RCX's

Grading:
    Determine the correct path (A).
    Successfully collect data (B).
    Show up and have fun (C).
    Bicker and argue with others (D).
    Don't show up to class (F).

3.2 The Structures Badge

The Structures badge will cover *jumps, loops*, and *forks*, which are the basic control structures in ROBOLAB. Combined these functions will allow you to create programs to make your robots react to their environment in an autonomous fashion.

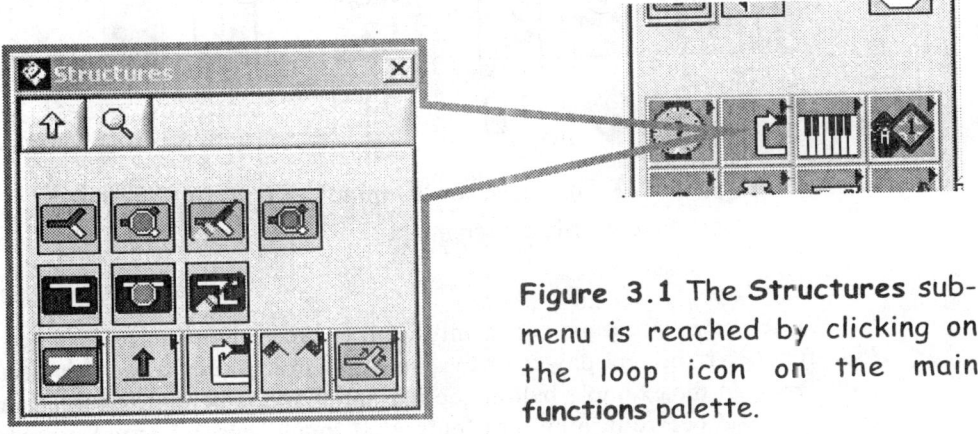

Figure 3.1 The **Structures** sub-menu is reached by clicking on the loop icon on the main **functions** palette.

3.2.1 Jumps

The *jumps* are equivalent to the *GoTo* commands in C and FORTRAN. When the program reaches an up arrow it "jumps" to the corresponding "land" down arrow. There are 5 color-coded jump/land and one generic jump/land pairs of arrows. *Jumps* are most commonly used for creating a simple loop, as shown below in Figure 3.3. However, it is usually better programming etiquette to use the *loop* functions rather than *jumps* because the jump command creates an infinite loop – meaning the program never ends (the stop light is never reached).

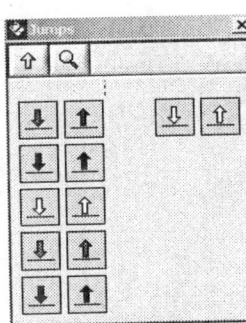

Figure 3.2 Jumps sub-menu

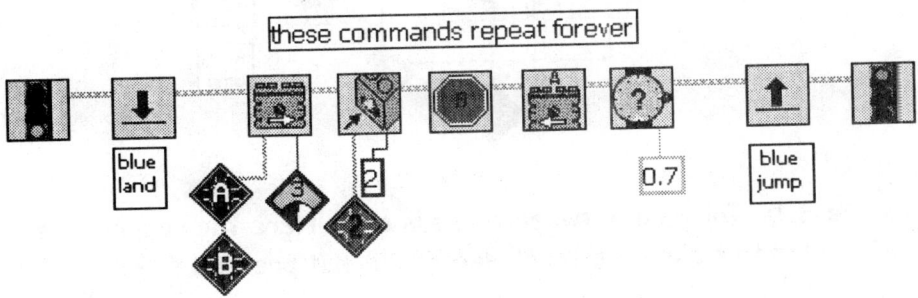

Figure 3.3. An infinite loop using the blue jump/land.

In addition to jumping backward in a program as in Figure 3.3, jumps can also be used to jump forward in a program, as shown in Figure 3.4. Here we've also used the generic jump function, which can be used to create up to 20 different jump/land pairs in case you want to use more than the 5 standard colored jumps. The jump number (#8 in this case) must be wired to both the jump and land functions.

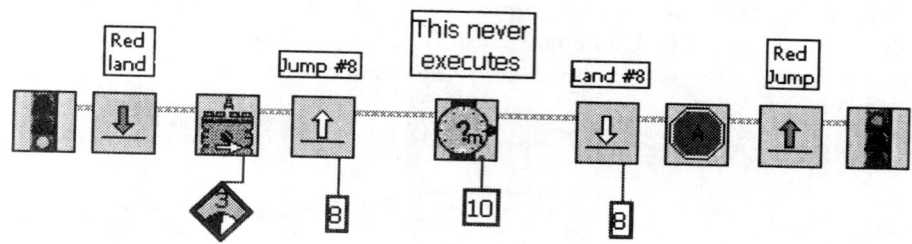

Figure 3.4 The wait for 10 minutes function is never executed in this program.

While *jumps* are most commonly used to create loops (which is better done with the *loop* function), the real power of the jump function comes when it is used in combination with *forks*. In the example below, the red jump/land is being used to ensure touch sensor 1 is not pressed before turning on Motor A. If touch sensor 1 is not pressed, then the upper portion of the fork executes. If the touch sensor is pressed, then Motor B is turned on until the touch sensor is released and then the program jumps from the lower fork to the upper fork. Here the *jump* is being used to perform a task that *loops* cannot. *Forks* will be covered in detail in section 3.2.3

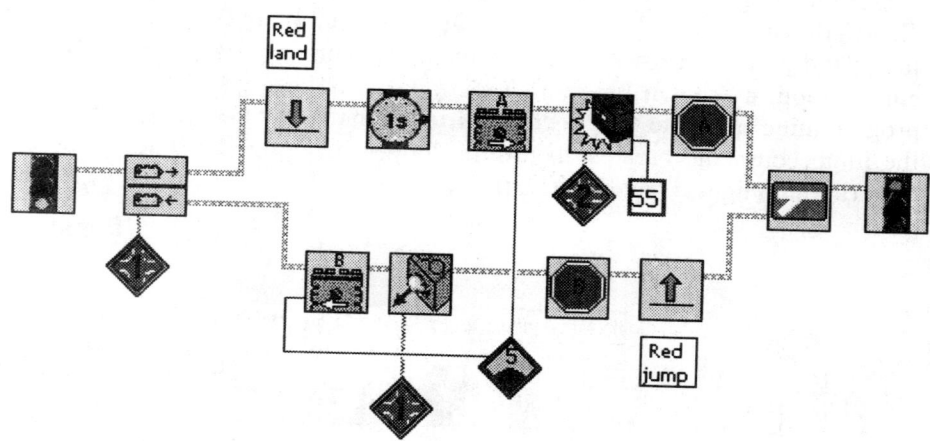

Figure 3.5. Jumping between forks is okay. Here the red jump is being used to make sure touch sensor 1 is not pressed.

 While using *jumps* and *forks* together is highly encouraged, *NEVER JUMP FROM ONE TASK TO ANOTHER! Tasks* will be covered in section 3.4.

Another really useful trick is to use multiple jumps of the same color, which works as long as you have only one land. In the figure below, we created a simple controller that uses the touch sensor to determine which direction Motor A will spin.

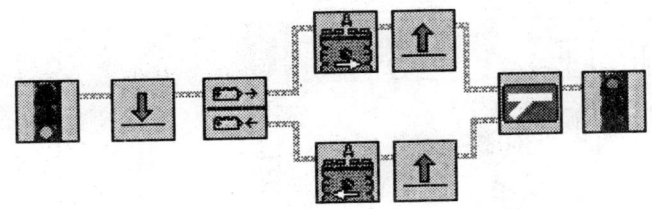

Figure 3.6. Multiple jumps of the same color are okay, as long as you have only one land.

3.2.2   Loops

Loops are conditionals which allow you to repeat specific sections of your program over and over. The **loop** sub-menu contains many types of *loops*. On the right side of the sub-menu is the *generic loop* function. On the left side of the sub-menu are over twenty *loops* that depend on sensor, container, timer, mail, clock, and camera values.

All the *loops* have the same basic form; loops are defined by one of the *start loop* commands and terminate with the *end loop* command. All of the functions that you want to repeat (loop) are bracketed by the *start* and *end loop* commands:

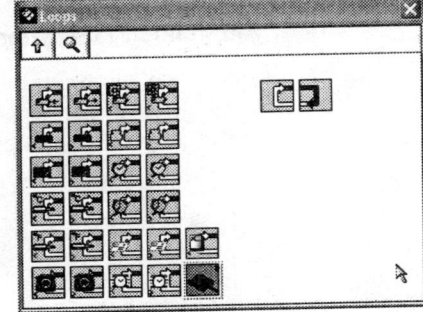

Figure 3.7. Loops sub-menu.

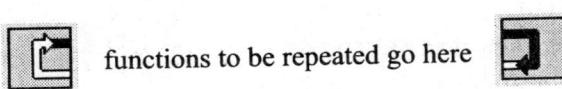

We saw the *generic loop* used in Figure 2.60 briefly. The *generic loop* is similar in function to a *jump* function, the exception being that you can specify the number of times the loop repeats. The number of repeats is wired to the *start loop* command.

| If no value is wired, the default is to loop only twice! |

The figure below illustrates a simple loop that plays musical D-note 5 times before turning the RCX off.

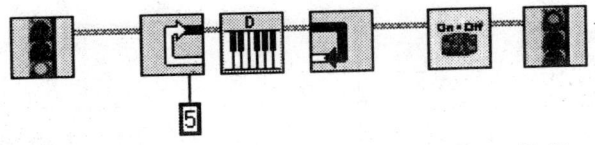

Figure 3.8. A simple loop that repeats 5 times.

The next figure shows the same set of functions used in Figure 3.3 sandwiched between the *start* and *end loop* commands. Unlike the *jump*, the *loop* will repeat a random number of times before the program ends.

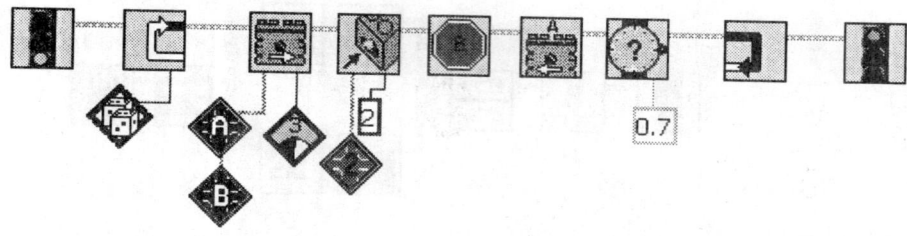

Figure 3.9. This program will loop (repeat) a random number times.

Unlike *jumps*, *loops* can be nested quite easily. Figure 3.10 shows an attempt to utilize nested jumps. Because the inner (blue) jump/land pair repeats indefinitely, the outer (red) jump is never reached. The program is "stuck" in the inner jump/land. This is why you shouldn't use *jumps* to do the work of a *loop*!

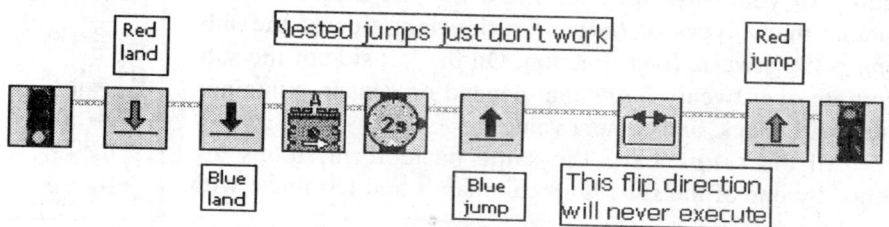

Figure 3.10. An unsuccessful attempt at nesting jumps.

In Figure 3.11 we have created a program that uses nested loops. The inner loop repeats twice, adding 5 to the red container each time. The outer loop repeats 5 times, subtracting 1 from the red container each time. The trick, however, is that the inner loops executes twice *each time* the outer loop is repeated. Thus, the inner loop repeats a total of 10 times.

If we were to keep track of the value of the red container it would look like this:

5, 10, 9, 14, 19, 18, 23, 28, 27, 32, 37, 36, 41, 46, and finally 45.

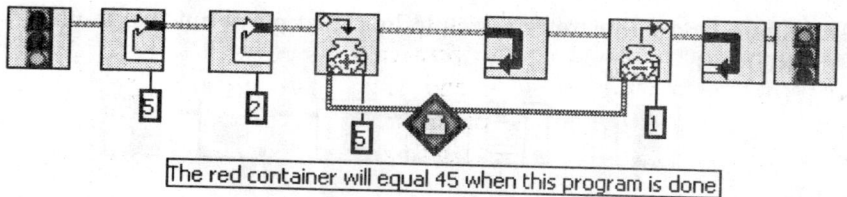

Figure 3.11. Example of nested loops.

In figure 3.12 we've created a program for a LEGO cockroach. The bug will move around randomly as long as it's dark (light sensor value less than 40). When the lights are turned on, the bug will run away at full power. Since we've used a *jump*, our program will repeat forever (or at least until the batteries die).

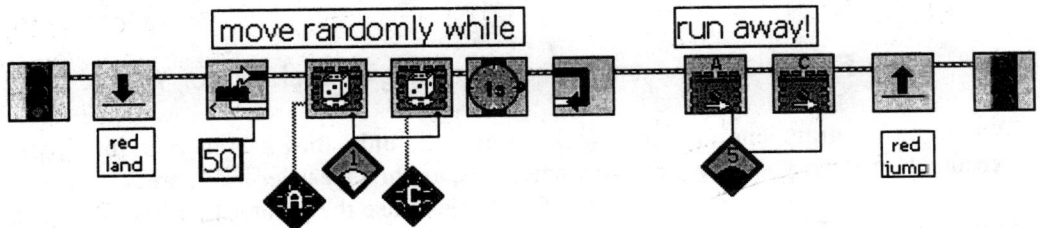

Figure 3.12. A program for a LEGO cockroach.

The following is an example of a LEGO alarm clock done with a pair of *loops*. The program begins by setting the RCX clock, which is displayed on the LCD panel in hours.minutes, equal to your computer's clock. The program then loops, checking the time every 0.5 minutes, until the clock exceeds 480 minutes, which is the same as 8:00. Then it plays sound #2 five times before ending. We could have used the *wait for clock* function instead of the clock loop. Can you modify this program to add a snooze for 10 minutes button? Give it a try!

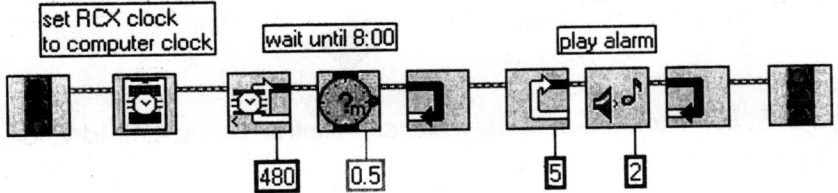

Figure 3.13. A LEGO alarm clock set for 8:00.

3.2.3 Forks

ROBOLAB *forks* are equivalent to the traditional *If-Then-Else* statements used in many programming languages and spreadsheets. When a fork is reached in the program, one of two "paths" will be taken. For example, in the figure below if the touch sensor is pressed Motor B will turn on. If touch sensor is not pressed, Motor A will turn on. All *fork* commands require that a *merge fork* command (the green and white icon) be used later in the program. The merge fork command is equivalent to the *end if* command in many text-based programming languages.

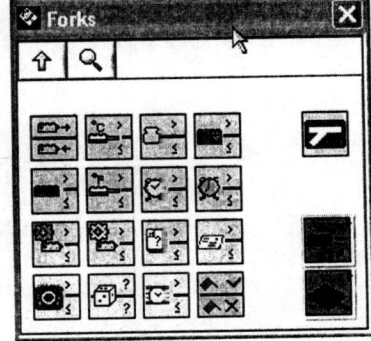

Figure 3.14. Forks sub-menu.

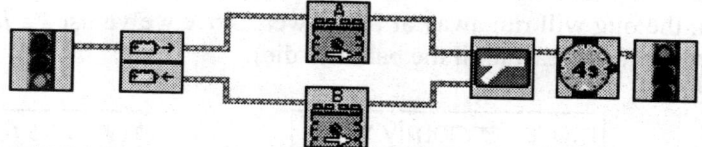

Figure 3.15. Sample program using a touch sensor **fork**.

When the program encounters a fork command, only one of the two fork paths will execute. However, you may jump between paths, as shown earlier in Figure 3.5.

All of the fork commands on the fork sub-menu use the greater-than/less-than condition to determine which path to execute. With the exceptions of the touch and the random forks, all fork commands require you to wire a threshold value, or decision level, to the command. For example, Figure 3.16 illustrates a light sensor fork with a threshold value of 43. If the light sensor value exceeds 43, then the upper path executes. Conversely, if the value is 43 or less, the lower path executes.

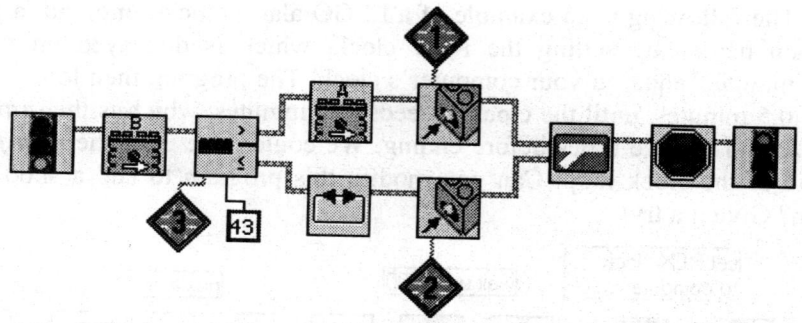

Figure 3.16. Example of a light sensor fork with threshold value of 43.

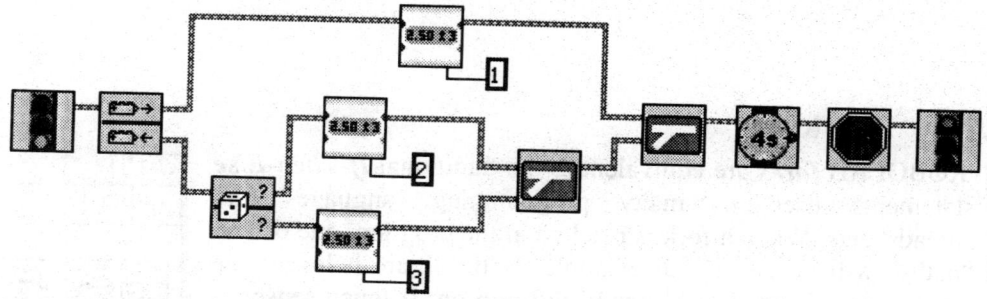

Figure 3.17. The touch and random forks do not require threshold values.

Unlike the other forks, neither the touch sensor nor the random fork commands require a threshold value to be used. In the example shown the number 1, 2, or 3 will be displayed on the LCD panel of the RCX for 4 seconds depending on which path is taken.

 **TIP** Displaying data to the LCD panel is a common way of debugging a program since it lets you see where you are in your program or what the container value is. In this example, you can determine which path was executed, depending on which value is displayed. Playing a sound is another common debugging technique.

Notice we've used nested forks in Figure 3.17 to achieve 3 possible paths. There is no limit to the number of nested forks that can be used. In the program below, we've used nested forks to divide the light sensor value into 4 different levels. Notice that all the forks need to be merged before the end of the program.

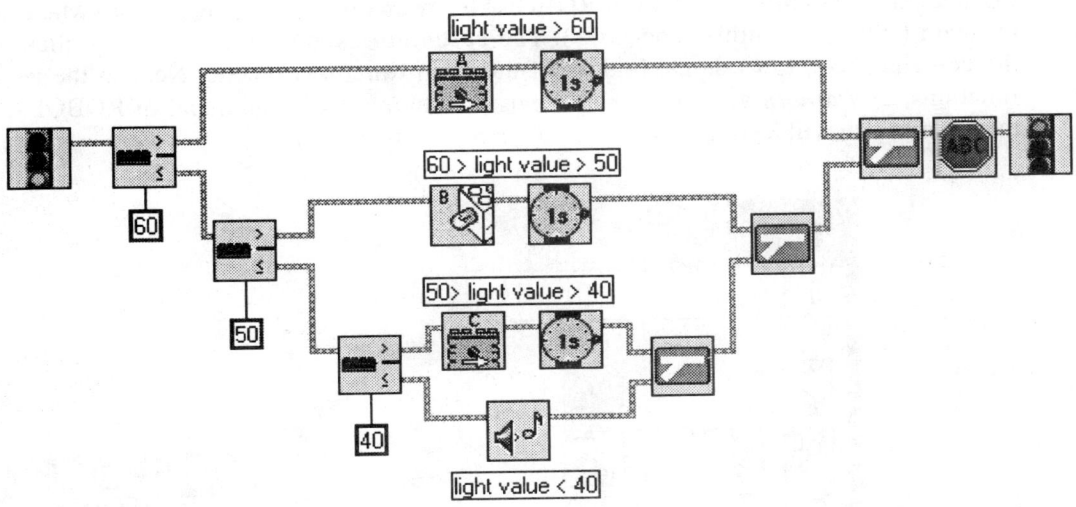

Figure 3.18. Example of nested light sensor forks.

The **Equal Forks** sub-menu can be accessed by the icon on the far right of the **Forks** sub-menu. The *equal forks* are identical to the standard *forks* except that the conditional statement is "equal to/not equal to" instead of "greater than/less than." Like the standard *forks*, there are *equal forks* for all the sensors, container, timer, clock and mail.

The program below initializes the touch sensor counter and then waits a random time, over and over until the touch sensor is pressed at least once. Then it plays the default sound and ends.

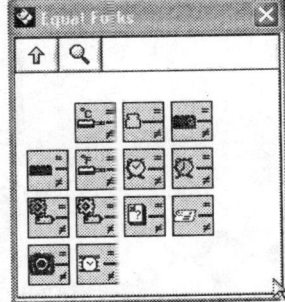

Figure 3.19. Equal Forks sub-menu.

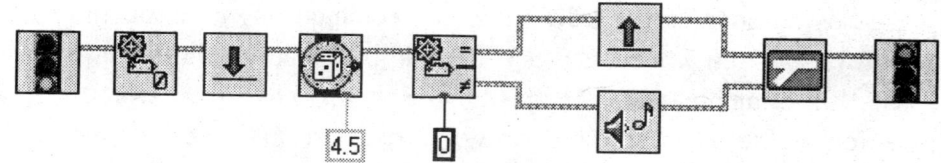

Figure 3.20. Example of a touch sensor equal fork.

3.3 The Containers Badge `Containers`

Containers are *Global Variables* in ROBOLAB, meaning they can be used anywhere in the program (in all subroutines and tasks). The *containers* submenu is access by clicking on the container icon towards the bottom of the main **functions** palette. Next to the *wait for* functions, *containers* are probably the most commonly used command in ROBOLAB. In this section you will learn how to use containers effectively.

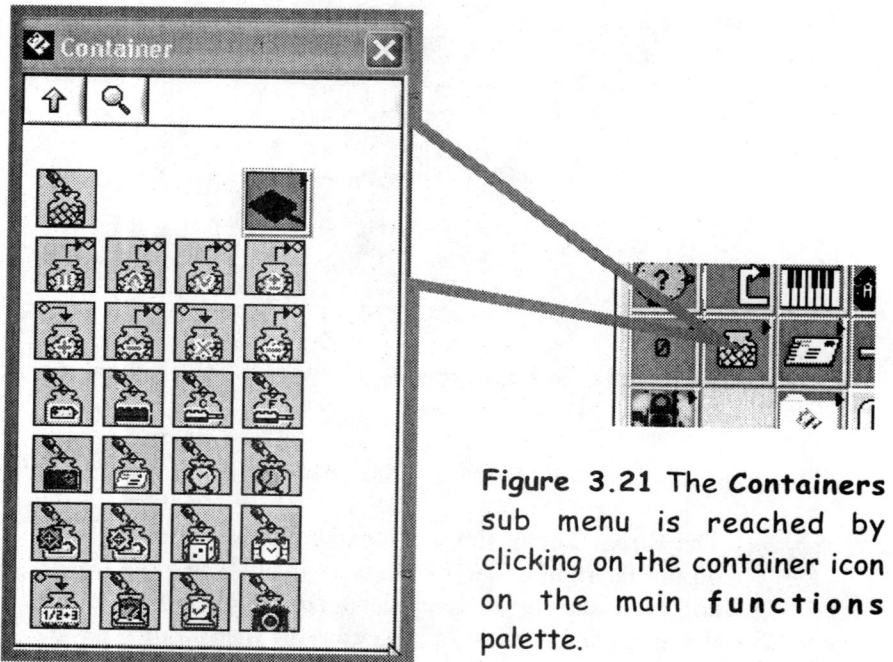

Figure 3.21 The **Containers** sub menu is reached by clicking on the container icon on the main **functions** palette.

3.3.1 Container Basics

There are 3 basic *containers* to use: red, blue and yellow. All of the *container* functions are found in the **Containers** sub-menu (Figure 3.21), but the containers themselves are found in the **Modifiers** sub-menu (Figure 2.49). Containers are used to store numbers. In ROBOLAB *all containers are integers*, which means that you cannot store floating point numbers; ROBOLAB will round down to the nearest integer (e.g. 1.9 becomes 1). In this example, all we are doing is setting the red container to 10.

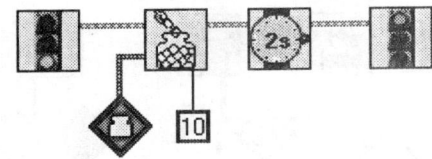

Figure 3.22. Set the red container to a value of 10.

Admittedly, this isn't very exciting. In order to do something useful, we have to introduce *container values*. Think of *containers* as the color of the jar (red, blue, yellow) and *container values* as the amount of stuff in the jar.

CONTAINER	CONTAINER VALUE
Containers use the brown wires. You wire containers either to container functions or to specify which container you are using in a fork or loop.	Container Values are used just like any numeric constant. They are wired with a blue wire.

In this example, we again set the red container to 10 and then we used the value of the container to make the program wait 10 seconds. Notice that *container values* are wired using a blue wire, just like a numeric constant, and *containers* are wired with a brown wire. Container values can be used anywhere a numeric constant is used.

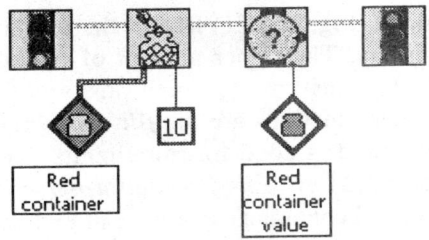

Figure 3.23. Using a container and a container value.

The purpose of the program in the next example is to take a light reading and then wait for the light to increase by a 5 units before playing some music. In order to accomplish this, we have used yet another modifier, the *port value*, to store the value of the light sensor on port 1 into the yellow *container*. We then add 5 to the yellow *container* and use this *container value* as the threshold for the *wait for light* function. When the light sensor value exceeds the value of the yellow container a quarter D-note is played.

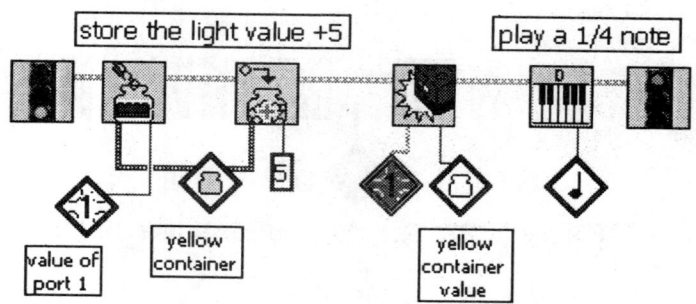

Figure 3.24. Using port values and container values.

Just like *container values, port values* can be used wherever a numeric constant is used. **Port values** return the value (integer) of the sensor attached to the corresponding port. For a light sensor, this would be the intensity of the light. For touch sensor, this is either 1 (pressed) or 0 (released).

As a side note, this type of program can be used to make an advanced line following robot. By starting with the light sensor over the black line and storing the sensor value to a container, the threshold value can be determined for almost any lighting condition. You don't have to re-program the threshold value for different ambient light conditions! This is commonly referred to as *calibration by demonstration*.

3.3.2 Container Sub-menu

The **container** sub-menu (Figure 3.21) has all the container functions, which are functions that operate on containers. The upper portion of the sub-menu has arithmetic and logic container functions. The arithmetic functions are fairly straightforward. The most commonly used container functions are the *fill container* and *add to container* commands. The *fill container* command is good for initializing a container to a specific value, as we did in Figures 3.22 and 3.23. The *add to container* command is good for counting (the default is to add 1 to the red container) as we did in Figure 3.11.

The logic functions are equivalent to the AND and OR logic gates which you will probably learn about in an introduction to electronic circuits course.

The lower portion of **containers** the sub-menu has the all the container functions relating to sensors, timers, clock, etc. All of these functions basically fill the specified container with the value of the sensor, timer, clock, etc. In each case the container is wired with a brown wire and the value is wired with a blue wire.

The following program measures the time, in tenths of a second, between two touch sensor presses and then displays the time on the LCD display of the RCX for 4 seconds. The program utilizes both the blue *timer* and the *timer value*. Just like the *container value* and *port value*, the *timer value* is used just like a numeric constant and is wired with a blue wire.

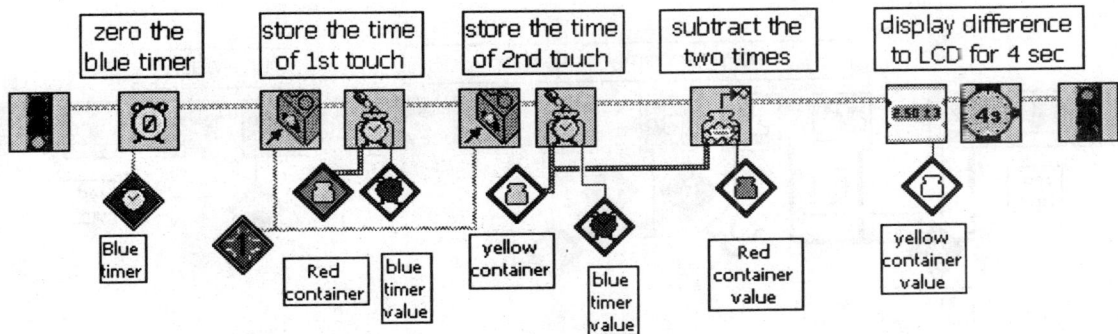

Figure 3.25. Measuring the time between touch sensor presses.

3.3.3 Container Wait For, Loop, and Fork functions

The *container wait for*, *container loop*, and *container fork* functions operate in the same fashion as the other *wait for*, *loop*, and *fork* functions described earlier. The *container wait for* causes the program to wait until the specified container is exactly equal to the integer value specified. The *container loop* will cause the loop to be repeated as long as the specified container is less than or greater than the specified value. The *container fork* is a conditional that will cause one of two program forks to be executed depending on whether the container is less than or greater than the specified value. For all the *container wait for*, *container loop*, and *container fork* functions, the container is specified by wiring one of the 3 colored containers (on the **modifiers** sub-menu) using a thick brown wire. The threshold value is specified by using a thin blue wire to connect a *numeric constant*, *port value*, *container value*, or *timer value* (all found on the **modifiers** sub-menu).

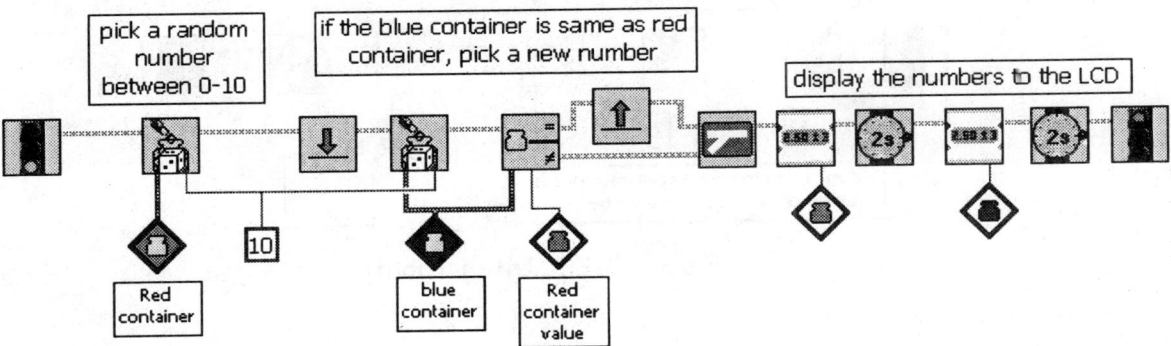

Figure 3.26. A program that selects two different random numbers.

In figure 3.26 we've created a program that will select two random numbers between 0 and 10. The first number is stored in the red container and the second is stored in the blue container. The container fork is used to ensure that the two numbers are not the same. If they are the same, the program jumps back and selects a new number for the blue container. Finally, both numbers are displayed to the LCD for 2 seconds each.

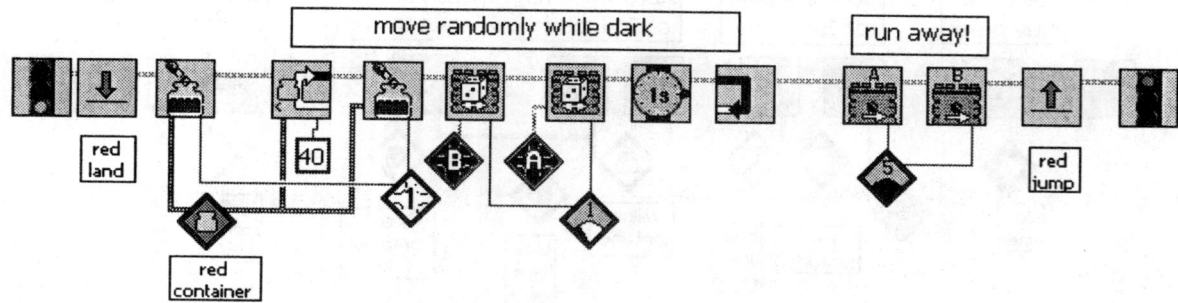

Figure 3.27. The LEGO cockroach program done with a container loop.

Here we've redone the LEGO cockroach program (see Figure 3.12) using a container loop instead of a light sensor loop. This also demonstrates a common phenomenon in programming: there is usually several ways to accomplish the same task using a different set of functions.

3.3.4 Integer Math

We keep stressing that the containers only work with integer values for a good reason. Realizing the consequences of integer math can save you a lot of headaches in the future. In integer math, all numbers are rounded down to the nearest integer. In the example below, 3 divided by 2 equals 1 in integer math. Thus, the end result is that "2" is displayed on the LCD panel. In general you always want to do your multiplications first and divisions last.

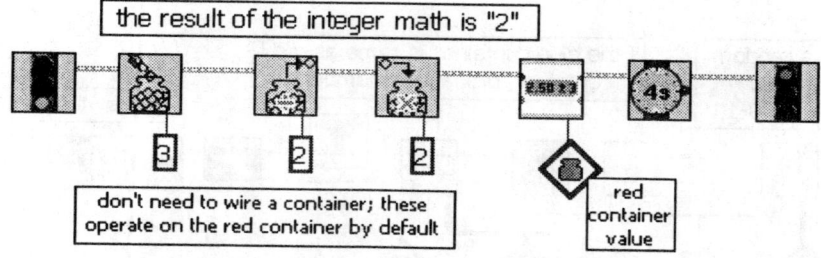

Figure 3.28. Integer math.

3.3.5 Generic Container

If you ever need to use more than 3 containers, you can use the *generic container* and *generic container value*, which allows you to use up to 31 different containers. You specify the container number using the *container.ctl* modifier, which is a pull-down menu listing the 31 container options.

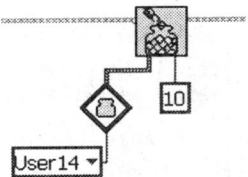

Figure 3.29. Example of filling generic container #14 with the integer 10.

3.3.6 Container Examples

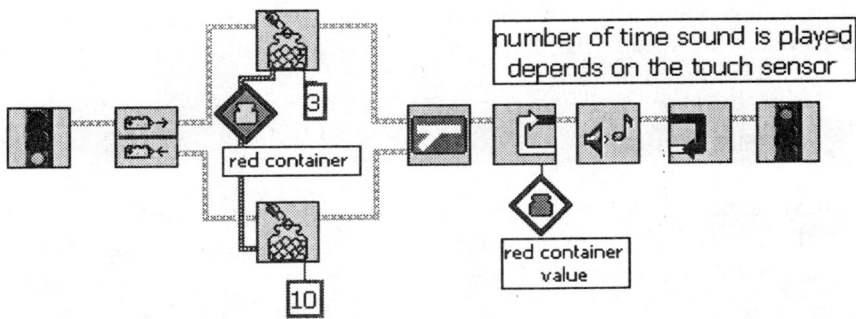

Figure 3.30. Example #1. In this example we've used the red container to specify the number of times the loop is repeated depending on whether or not the touch sensor is pressed.

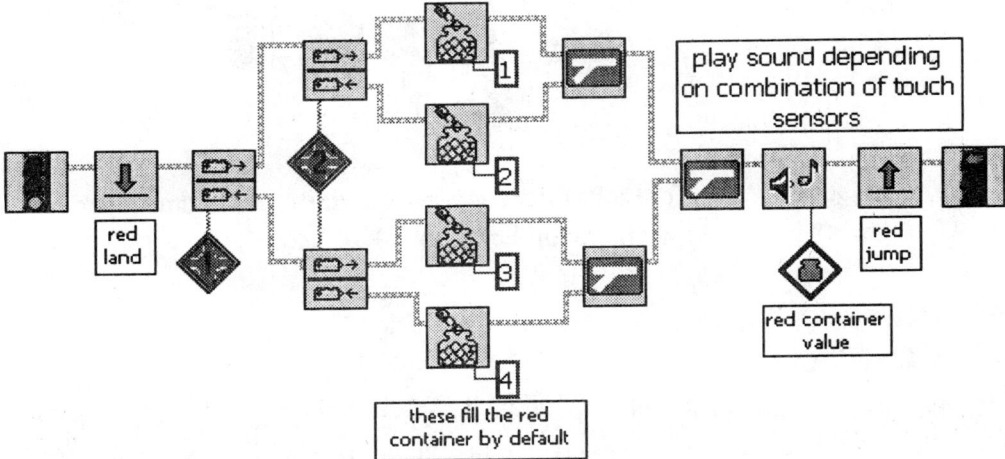

Figure 3.31. Example #2. Here we've created a simple musical instrument. Depending on the combination of the two touch sensors, we will play one of 4 system sounds.

3.4 The Tasks Badge   [Tasks]

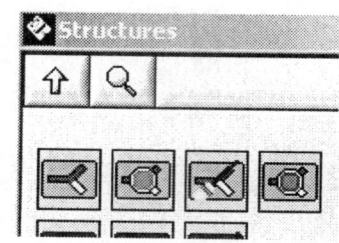

The RCX has the ability to multi-task, meaning it can run more than one program at a time. The way this is accomplished is with Tasks.

If you have ever surfed the Internet, you probably have encountered a web link which opens a whole new browser window when selected. You then have two browser windows open and closing one does not close both windows. ROBOLAB *Tasks* are similar. You can have your program spawn a new task, just like opening a new browser window, which will execute independently of the original task. *The program does not end until all the tasks have ended.* This means that every tasks must end with it's own red stop light. You can have up to 11 tasks running at the same time.

3.4.1   Task Splits

To spawn a new task, the task split command is used. This command allows you to essentially run two different programs at the same time. This is very handy for monitoring two sensors at the same time. Figure 3.30 shows a program which utilizes a task split to *independently* control Motors A and B and monitor the touch sensor on input port 1 and the light sensor on input port 2. Notice that because the tasks are independent, even if one motor stops (its task ends), the other will keep running.

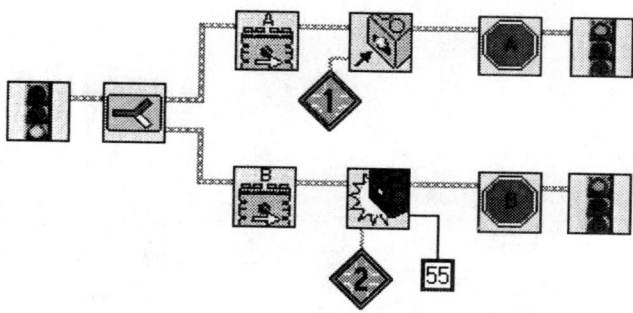

Figure 3.32. Task splits create independent programs that must both end with a stop light.

Hardware Conflicts

The ability to run independent programs is a double edged sword because the programs share the same hardware resources. If one task tells the Motor A to run forward and the other tells it to run in reverse at the same time, it's hard to predict what will happen. Conflict resolution will be covered in Chapter 5, but it is easy to avoid conflicting tasks by making sure tasks are not fighting over the same resource at the exactly the same time.

In the example below, the lower task will turn on Motor A in the forward direction. After 2 seconds the upper task will reverse the direction of the motor and end (even though the upper task has ended, the program will continue to run since the lower tasks is still active). Two seconds later (4 seconds into the program) the lower task will again reverse the motor direction (forward direction) and wait for 10 more seconds before stopping and ending the program. Since both tasks were controlling Motor A at different times, there was no conflict between tasks. The point is: make sure you are aware that hardware conflicts may occur with multi-tasking.

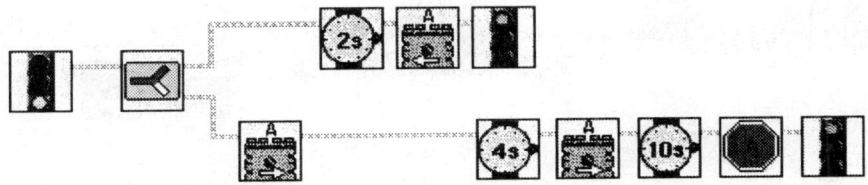

Figure 3.33. Watch out! Both tasks are trying to control motor A.

In this next example, the two tasks are fighting over the control of Motor A because they are both trying to control it <u>at the same time</u>. As we will describe in detail in Chapter 5, the upper task will win because it has a higher priority. Thus, the end result is that Motor A will run in the forward direction for 2 seconds. The lower task essentially does nothing since it lost out in the conflict over Motor A.

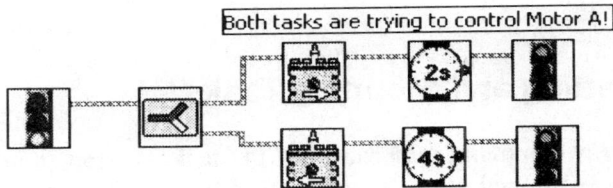

Figure 3.34. Task conflict over Motor A.

While controlling outputs may cause conflicts, tasks can share inputs without any problems. In the following program a single touch sensor is used to terminate both tasks. Both tasks are monitoring the same touch sensor at exactly the same time.

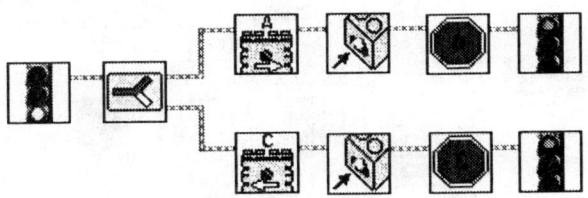

Figure 3.35. Two tasks monitoring the same sensor is not a problem.

A couple of quick reminders about tasks:
- <u>You cannot merge tasks.</u> You can only merge forks.
- A task split is basically the same thing as having another program running.
- If you use a task split, each separate program MUST have its own stoplight.
- Be aware of hardware conflicts when two tasks try to control a single output at the same time.

o   NEVER use a *jump* to jump between tasks.
o   NEVER use a *jump* to jump from one side of a task split to the other side.

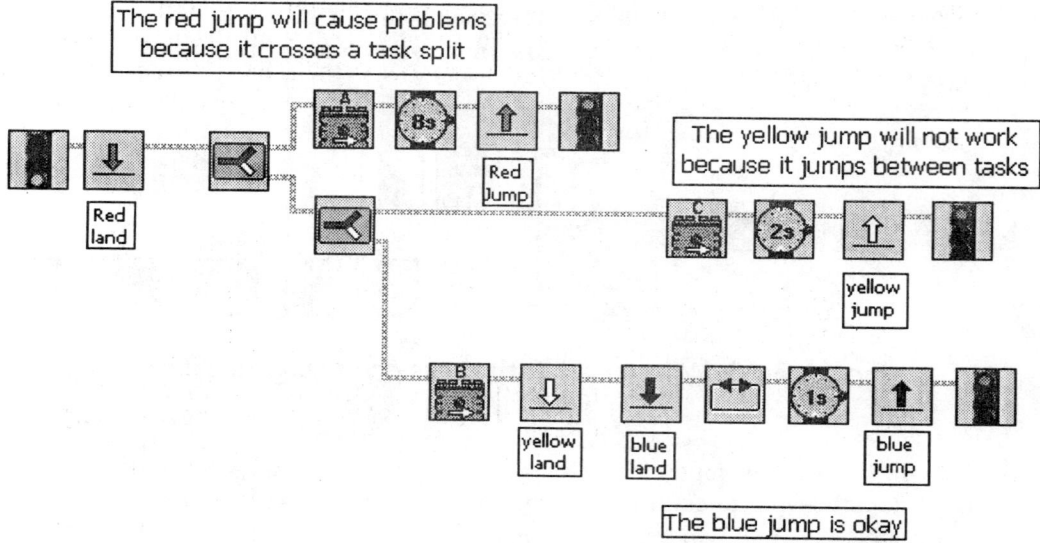

Figure 3.36. Bad (illegal) jumps used in combination with task splits.

3.4.2   Starting and Stopping Tasks

 The *stop tasks* command will cause all the tasks to stop running. In the program below, both tasks are stopped after 10 seconds. The lower task may finish executing before 10 seconds has elapsed, but this does not affect the upper task. Note that the default sound is never played since the *stop tasks* command is executed first, which essentially ends the program.

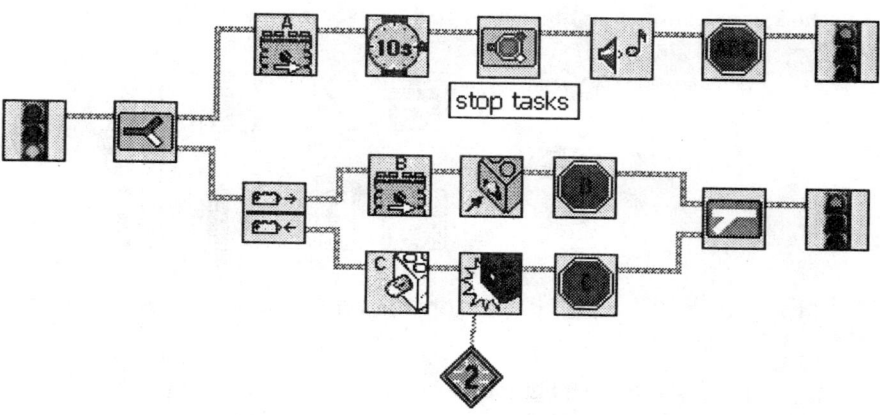

Figure 3.37. Both tasks will stop after 10 seconds.

In addition to stopping tasks, it is also possible to restart tasks using the *start tasks* command. In the example below the upper task will cause both tasks to stop if more than 10 seconds has elapsed. However, if the touch sensor is pressed before 10 seconds has elapsed, the lower task will stop Motor B, send the number 10 out as mail via the IR port, and then restart both tasks to from the original task split. The *start tasks* command essentially restarts the upper task (which happens to be task 1) from the task split. The program will "stay alive" as long as the touch sensor is pressed at least once every 10 seconds.

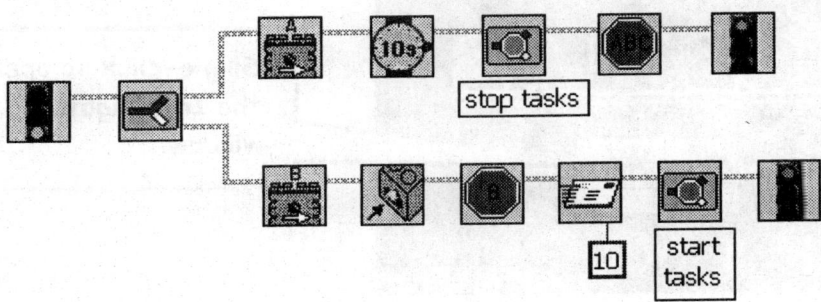

Figure 3.38. Tasks will restart from the tasks split.

3.5 Investigator Basics

Welcome to Investigator, the third and last programming mode in ROBOLAB. The main new feature in Investigator is the ability to collect sensor data using the RCX. In this section we'll cover some the basics of getting around.

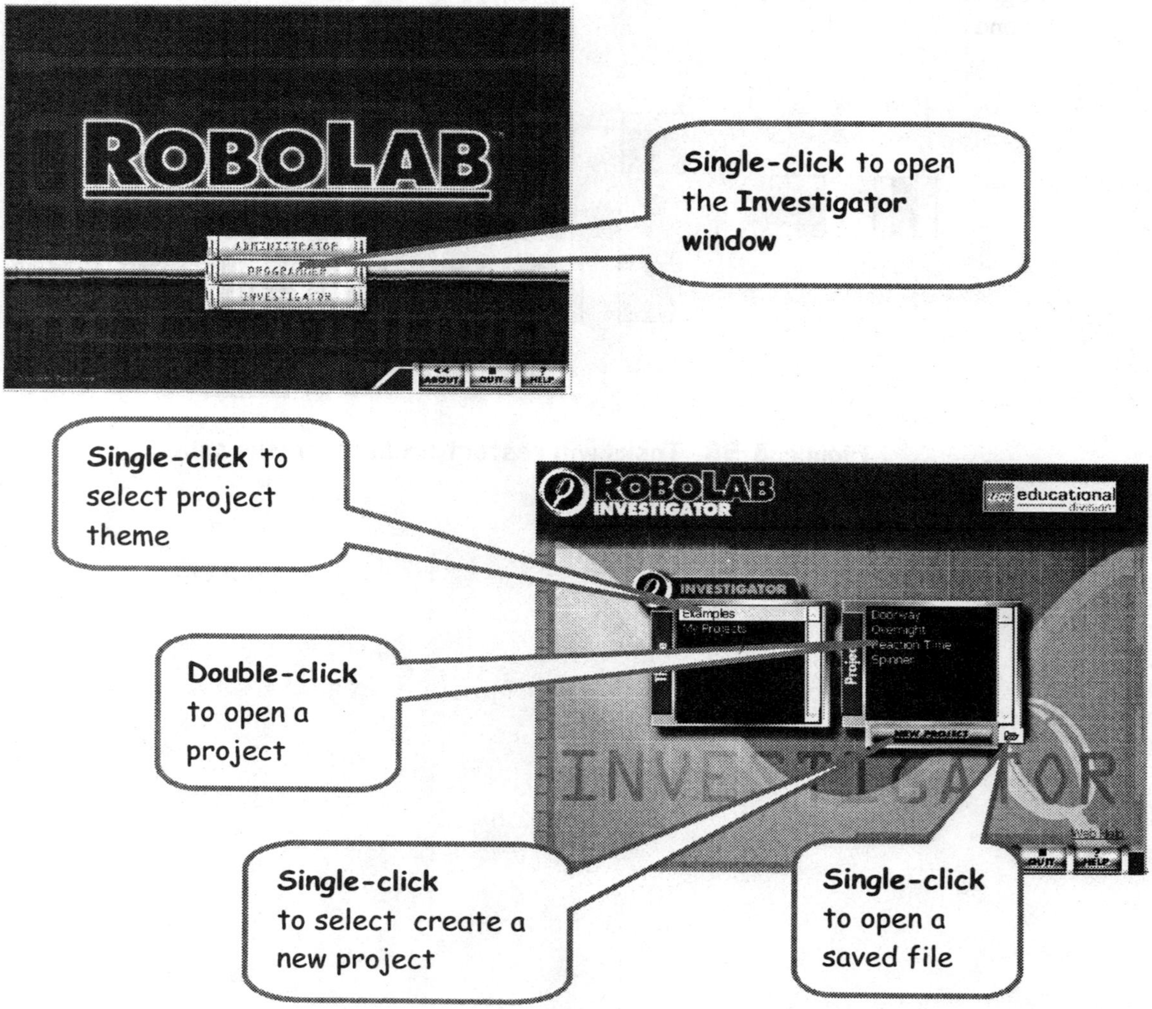

Single-click to open the **Investigator** window

Single-click to select project theme

Double-click to open a project

Single-click to select create a new project

Single-click to open a saved file

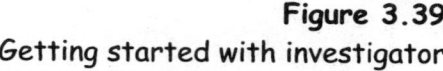
Figure 3.39
Getting started with investigator

 The open saved file option is only available in ROBOLAB version 2.5.1 and higher.

When you open an investigator project, two windows open: the **Navigation window** (left) and the **Project Working Area** (right).

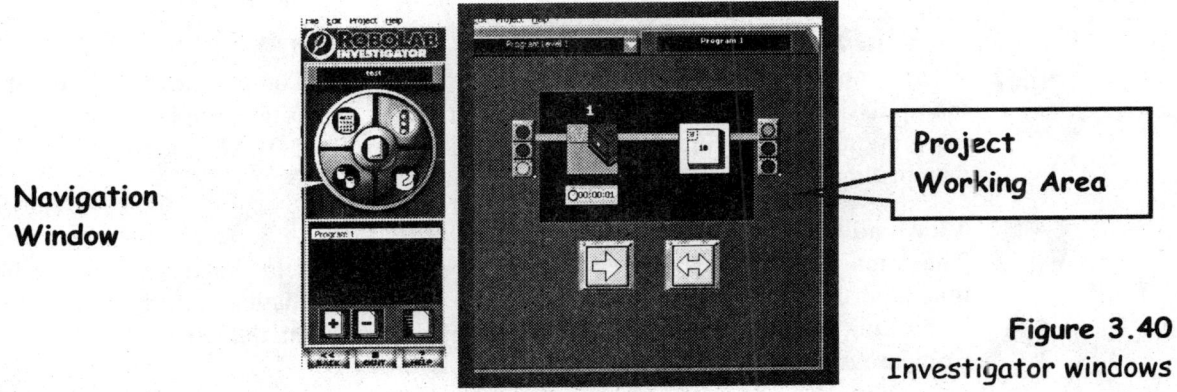

Figure 3.40
Investigator windows

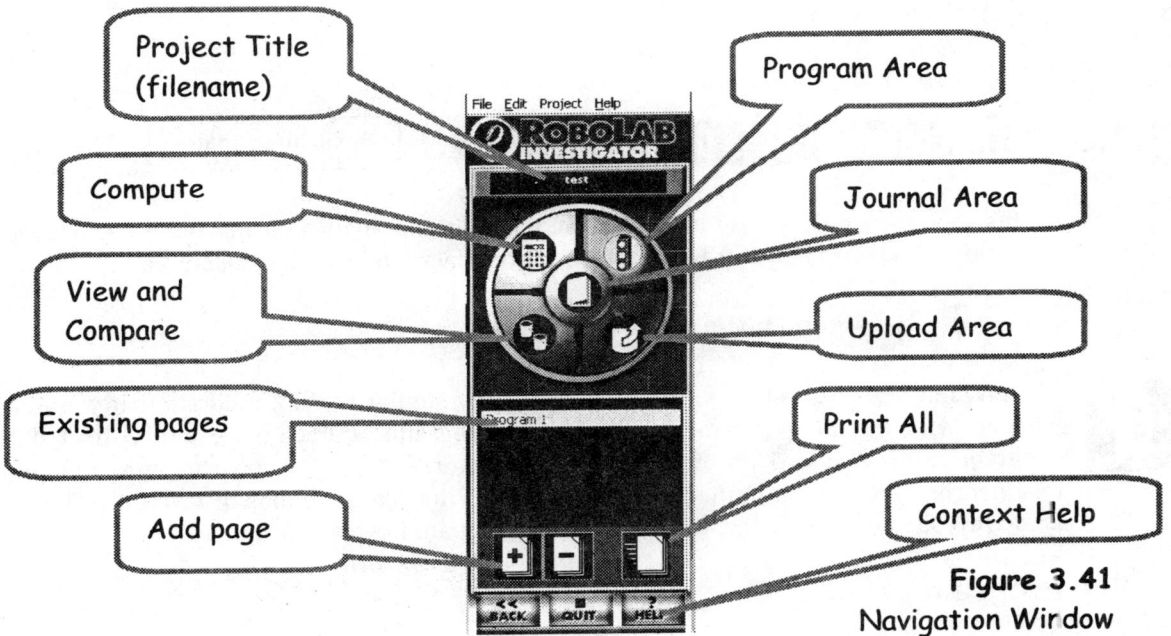

Figure 3.41
Navigation Window

Unlike Pilot and Inventor, Investigator has 5 different **project areas**, which are accessed using the **Navigation Window**. The **Navigator Window** always stays open. The **Project Working Area** will change as you move between the different project areas.

Program Area:
The Program Area is very similar to Pilot and Inventor modes. You write your ROBOLAB programs here. You also download your programs to the RCX in the Program Area. What's new in Investigator is the ability to collect sensor data using the RCX's built-in data acquisition system.

Upload Area:
Once you've collected some data using the RCX, you can upload the data back to the PC. Each data set must be uploaded to a new Page in the Upload Area.

 Compute Area:
The Compute Area is used to manipulate a data set. Computations can be simple manipulations such as addition and subtraction or more complex ones such as integration and differentiation. You can also write ROBOLAB code to process the data in almost any way imaginable.

 View and Compare Area:
The View and Compare Area is used to view data sets that have already been uploaded or manipulated. You can also get simple statistical information such as mean and standard deviations. You can also print your data from the View and Compare Area.

 Journal Area:
The last area, the Journal Area, is used to document your project. You can create simple reports or make notes for yourself. You can also import digital images to help you document your project.

3.6 The Basic Investigator Badge

In this section, we'll discuss basic data acquisition and analysis using the RCX. Using a Pilot mode type program, you'll learn the basics of data acquisition and analysis.

3.6.1 Program Area

 The program area has 5 levels. Levels 1-3 are very similar to Pilot mode in that you don't have to wire anything together and all your programs will compile and collect data. Program levels 4 and 5 are equivalent to Inventor in that you have to wire functions together, but you are not limited to the sequential Pilot-like programs of levels 1-3. For the Basic Investigator skill badge, we'll only cover Program Levels 1, 2, and 3.

Program Level 3

Just as we did in Pilot mode, we're going to skip Program Levels 1 and 2. We have full confidence that you can handle Program Level 3 since you've already earned the Basic Pilot and Basic Inventor skill badges back in Chapter 2.

To change the Program Level, click on the pull-down menu at the top left of the Program Area window. Select Program Level 3 from the list. Program Level 3 is very similar to Pilot mode with the added benefit that you can collect sensor data while controlling the outputs. Figure 3.42 shows a typical program for Level 3.

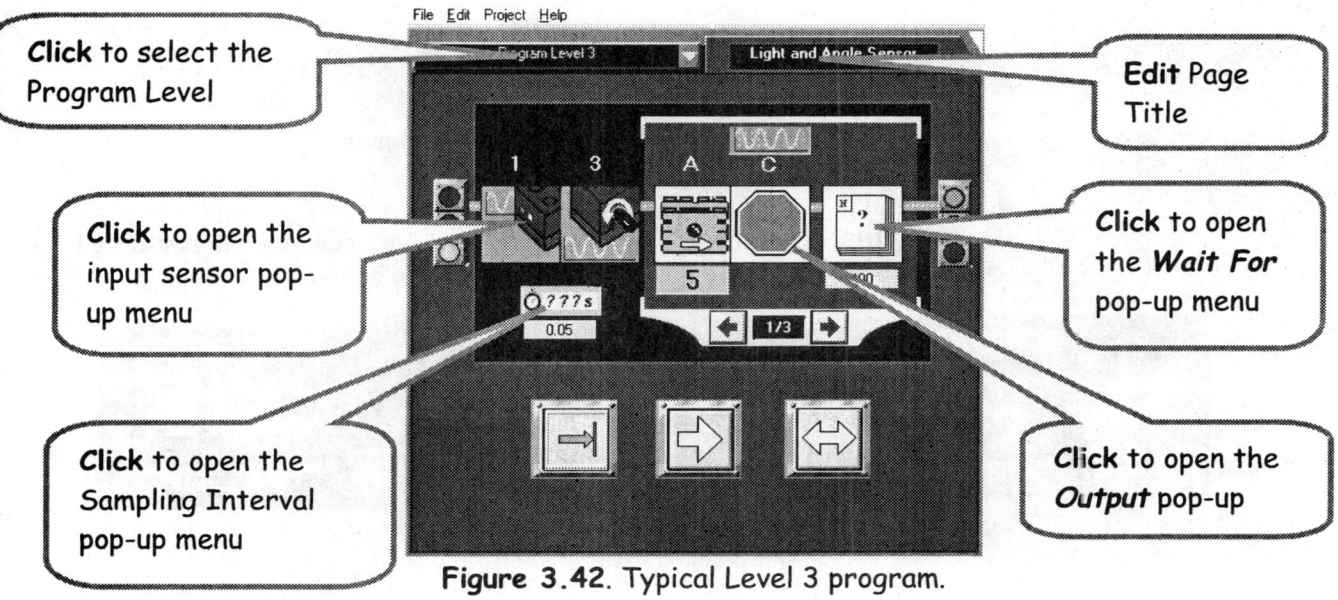

Figure 3.42. Typical Level 3 program.

The program starts by defining which 2 sensors will be used for data collection. While you cannot change the input ports (you have to use ports 1 and 3), you can change which sensors you want to use by clicking on the sensor's icon. The pop-up menus for input ports 1 and 3 are slightly different, as shown in Figure 3.43. On port 1 you can collect timer and camera data in addition to all the other sensors. On port 3 there is the option to not collect any data.

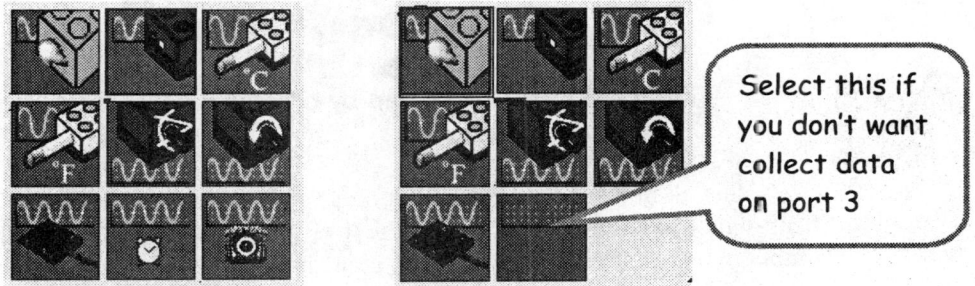

Figure 3.43. Sensor pop-up menus for input port 1 (left) and input port 3 (right).

Below the sensor icons is the **sampling interval** indicator. The **sampling interval** defines how often you want to collect data. Clicking on the indicator will open the **sampling interval** pop-up menu. The 5 **sampling interval** options are: 1 second, 1 minute, 1 hour, any time in seconds, or on touch sensor press. The last option will collect a data point for each of the sensors specified each time touch sensor 2 is pressed (this is why you can't collect data on port 2 – it's used for the touch sensor).

Figure 3.44. Sampling interval pop-up menu.

Just like in Pilot mode, you can have multiple steps. You can scroll forward and backward using the arrows on either side of the step number at the bottom of the frame.

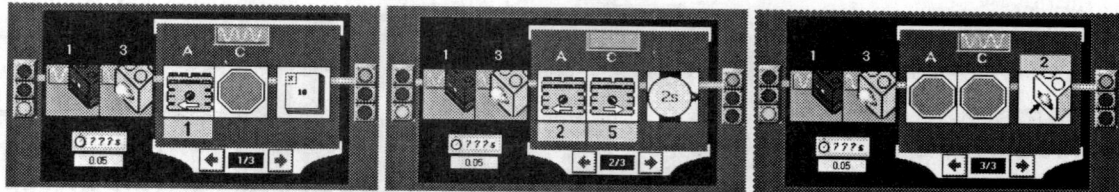

Figure 3.45. All Level 3 programs have 3 steps.

Unlike Pilot mode, however, you cannot add or subtract steps. **All Level 3 programs have 3 steps**. However, you can essentially remove a step by having the step do nothing for zero seconds as shown in Figure 3.46.

Figure 3.46. Skip a step by creating a step that does nothing.

Just as in Pilot mode, clicking on an output icon will open the **output** pop-up menu, from which you can select one of the 4 output options. You've probably noticed that the output ports cannot be modified. You are restricted to using **output ports** A and C for now.

Figure 3.47. Output pop-up menu.

Unlike Pilot mode, you have the unique option being able to set the Motor/Lamp power to be proportional to the sensor reading on input port 1. Clicking on the power level modifier just below the output functions opens the power level modifier pop-up menu as shown in Figure 3.48.

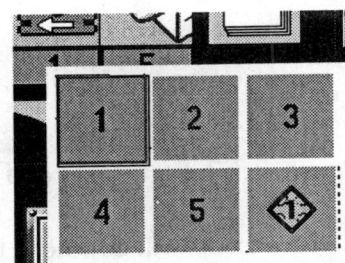

Figure 3.48. Motor power can be set for levels 1-5 or based on sensor port 1.

The last function (far left) in each step is the familiar *wait for* function. We can wait for touch, light, temperature, angle, or rotation just as before, but now we can also wait for a specified number of data points (top row in Figure 3.49). Exactly how long this will take will depend on the **sampling interval** you selected.

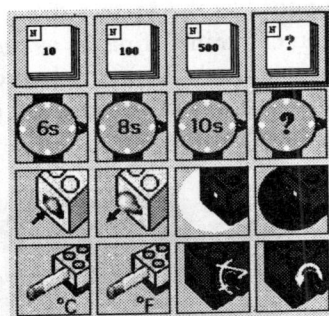

Figure 3.49. Wait for pop-up menu.

Figure 3.50. The Data Log pop-up menu lets you select either collect data (left) or don't collect data (right).

The yellow sine-wave icon above **output C** is the **data log** icon and it indicates whether or not you want to collect data during the current step. Clicking on the **data log** icon will allow you to turn on and off data logging for the current step.

Figure 3.51. Run mode.

The pink run-mode icon is also back from Pilot mode. Selecting *run continuously* will cause the program to repeat over and over indefinitely.

Figure 3.52. Run (left) and Direct Mode (right).

The large white **Run** arrow will download the program from the PC to the RCX. The double white arrow is used for running in **direct mode**. In **direct mode** the PC downloads the program to the RCX and immediately starts the program (no need to push the green run button). In **direct mode**, the RCX sends the data collected as fast as it can directly back to the PC via the IR port. The PC will display the data as it is received. In order for this to work, the RCX must remain within range of the IR tower at all times (a few feet typically). **Direct mode** can be very useful for debugging because you can see exactly what the data collected looks like. Note, this does not upload the data collected into a bucket; you must use the **Upload Area** to do that.

Finally, Investigator has the feature that you can create multiple *pages* in each area for a given project. You can think of pages as analogous to worksheets in Microsoft Excel. In the **Program Area**, you can write multiple programs, storing each one on it's own page, without leaving the current project. You can add or delete pages using the "+" and "-" icons at the bottom of **Navigation window**.

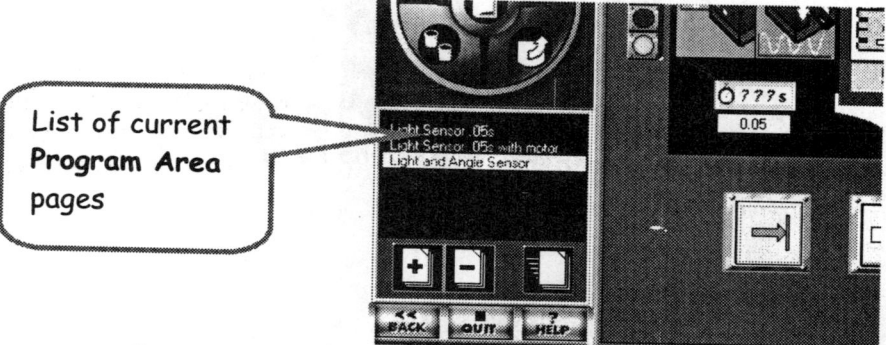

Figure 3.53. Pages allow you store multiple programs for a single project.

3.6.2 Upload Area

Once you have collected the data, it is stored in the RCX and must be uploaded to the PC in order for you to view, manipulate, and save the data. On the PC side, data is stored in one of 10 different colored buckets. Each page in the **upload area** can only have one bucket, so you are effectively limited to only 10 data sets.

While it may not be apparent, all the **upload area** pages share the same buckets so you should be careful not to use the same bucket for more than one page. You can change the name of a page by entering the name in the upper right hand corner, as shown in Figure 3.54.

Clicking on the large white arrow will upload the data from the RCX to your PC. The IR tower and IR port on the RCX must be facing each other and be in close proximity (just like when downloading a program). If you collected data for 2 sensors, the data for each sensor will be uploaded to a different page.

The 5 **Investigator Areas** do not share pages. That is, **Program Area** pages are independent from the **Upload Area** pages.

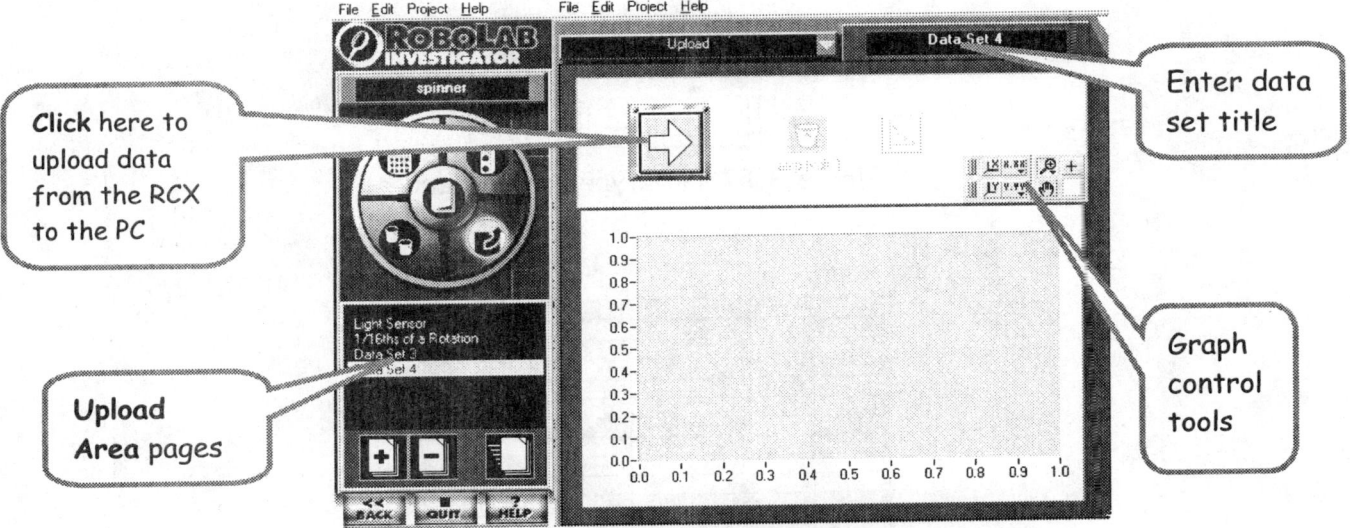

Figure 3.54. Upload Area.

All data is initially uploaded into the red bucket. If you are going to upload more than one data set, you will need to change the bucket color by clicking on the bucket icon *after the data has been uploaded*. This will open the **bucket** pop-up menu (figure 3.55). You can also change the names of the buckets by clicking on the title bar just below the bucket icon, which will open the **set titles** pop-up menu (figure 3.56).

Figure 3.55. Bucket pop-up menu.

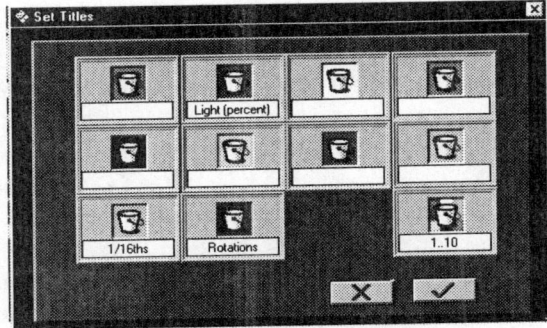

Figure 3.56. Set Titles pop-up menu.

The type of plot can also be controlled. Clicking on the icon just to the right of the bucket will open the **plot type** pop-up menu (figure 3.57). This menu has 4 plot types and a numerical value option. The numerical value option lets you see the raw data in 3 columns: data point number, time, and value (see Figure 3.58).

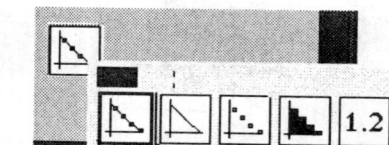

Figure 3.57. Plot Type pop-up menu.

Figure 3.58. The Numerical value Plot Type option lets you view the raw data.

Once you've uploaded a data set, you can adjust the way the graph looks using the graph control tools.

 Autoscale axis. This will adjust the X or Y axis ranges to fit the data to the screen.

 Lock autoscaling. Setting the switch to the right will lock the autoscale feature on.

 Format axis. Allows you to adjust the format of the axes.

 Zoom button. Allows you to adjust the zoom in several ways.

 Pan button. Allows you to move the graph around.

Enlarge button. Opens a full-screen version of the graph.

On the enlarged graph there are 2 cursors that can be moved around. The X and Y locations of the cursors are indicated at the bottom of the graph. You can also lock a cursor to the data set.

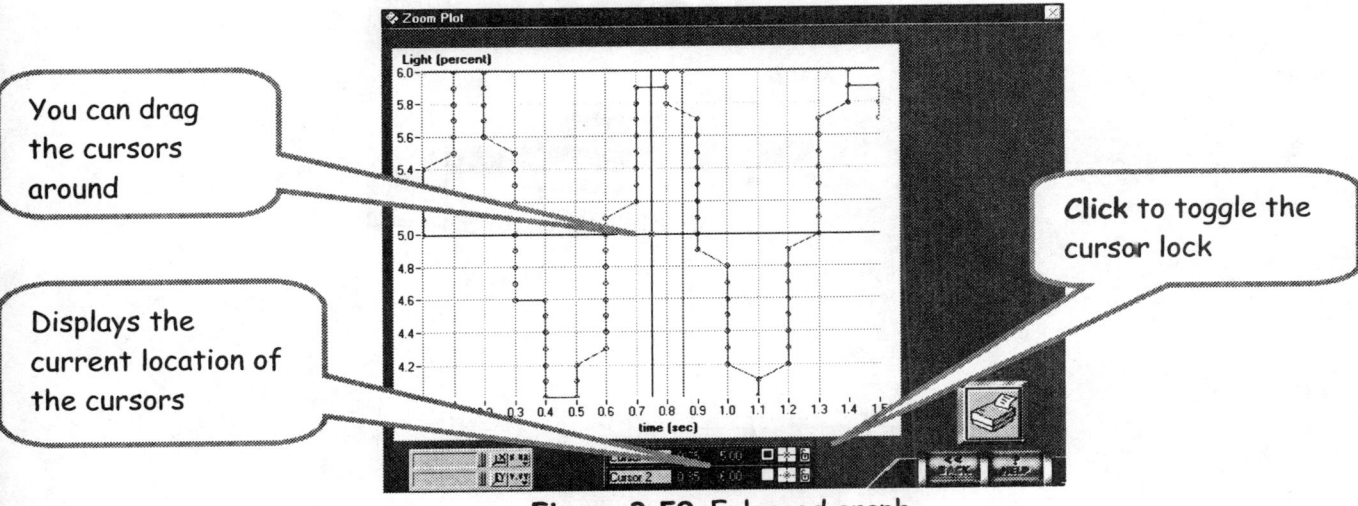

Figure 3.59. Enlarged graph.

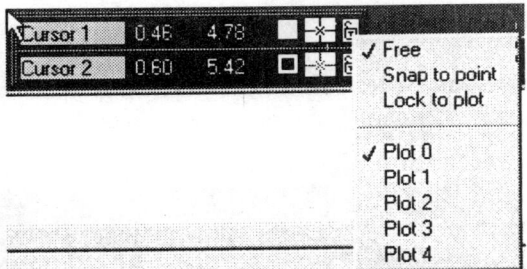

Figure 3.60. Lock Cursor pop-up menu.

Internet Upload

Internet upload allows you upload data over the Internet. The host computer (where you are getting the data from) must be set up as a ROBOLAB Internet Server (see chapter 4.6).

Exporting Data

If you prefer to process and/or graph the data in another program (e.g. Microsoft Excel), data collected can be exported to a file. To export data to file, select the **file** menu on the project window and scroll down to the Export: Page selection. A standard dialog box will open where you can type in the filename (typically with the .txt extension) and choose the location (folder) to save the file.

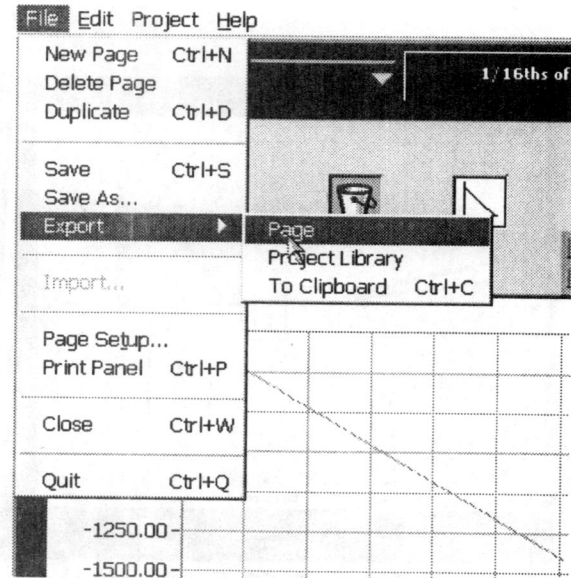

Figure 3.61. Use the File: Export: Page function to exporting data to a file.

3.6.3 View and Compare Area

 The **View and Compare Area** is used to view, compare, measure, and print data sets. Each page in the View and Compare Area can be used to perform one of these four functions.

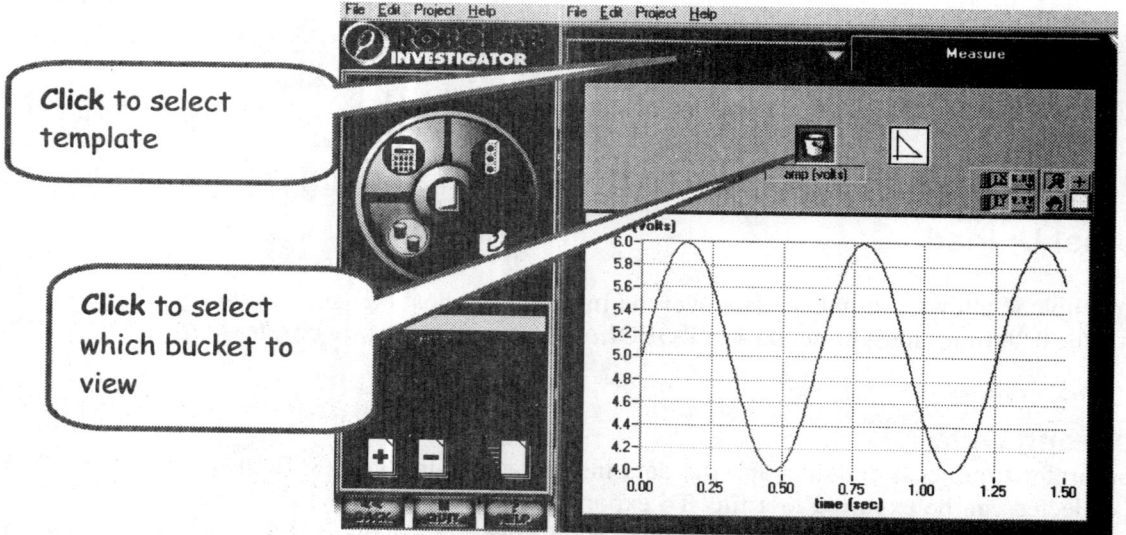

Figure 3.62. View and Compare Area.

The **view** template (figure 3.62) is used to view a single data set much like the Upload Area. You can select which bucket to view, the plot type and access the graph controls. In the **Upload Area**, the bucket icon defines which bucket the data will be stored in. In the **View and Compare Area**, the bucket indicates which data set is to be examined.

The **Compare** template is used to examine two data sets on the same graph. Just as before, you can control the plot type and access the graph control tools.

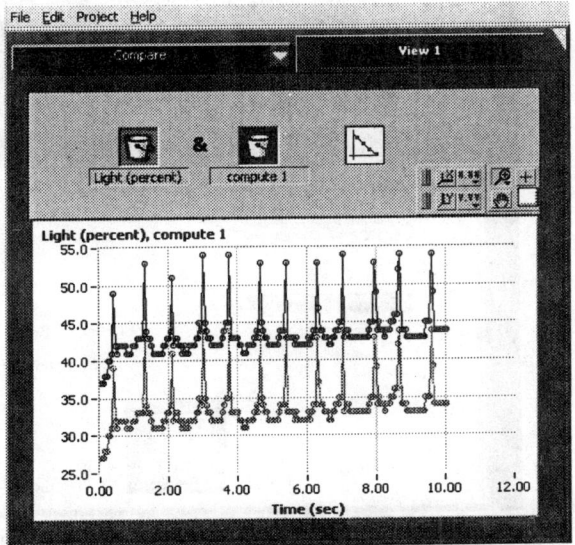

Figure 3.63. Comparing two data sets.

The **Measure** template (figure 3.63) can be used to perform basic statistical measurements on a single data set. You can determine minimum, maximum, mean, standard deviation, slope, and area under the curve.

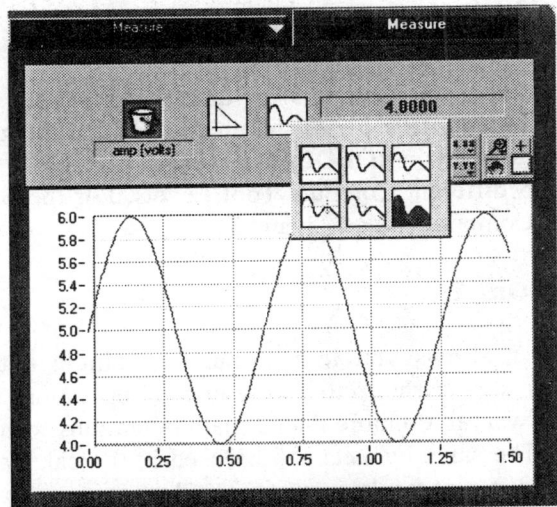

Figure 3.64. The measure template allows you to do simple statistics on data.

Finally, the **Print** template can be used to print one or more of the pages you've created in the View and Compare Area.

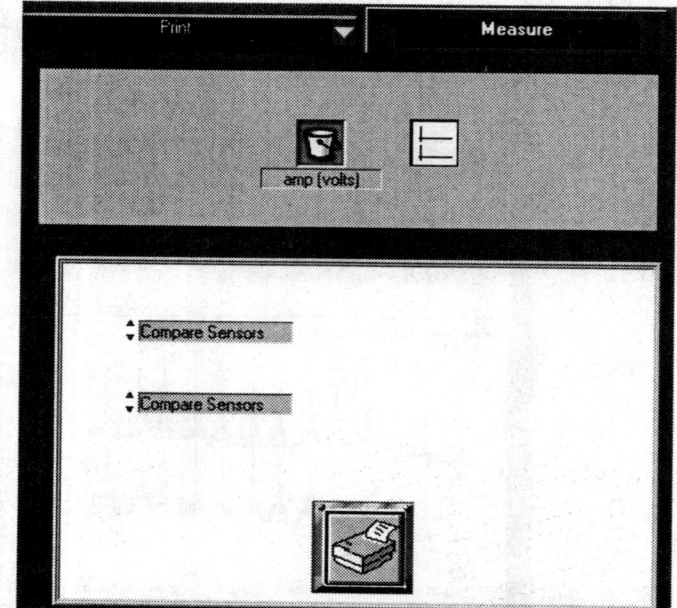

Figure 3.65. The Print template is used to print pages.

Exporting Data

Just as in the upload area, data can be exported to text file by selecting **Export: Page** from the **File** menu (see Figure 3.61). In the compare template, the exported text file will contain both sets of data.

3.6.4 Compute Area

The **Compute Area** is for manipulating data that has been uploaded into a bucket. For example, you may want to subtract the mean and then integrate a data set, which can be accomplished quite easily in ROBOLAB.

There are 5 different Compute Tool Levels. For the Basic Investigator skill badge, we'll only cover Compute Tools 1, 2, and 3.

Compute Tools 1

Compute Tools 1 allows you to manipulate a single data set using basic algebraic operations. You can perform up to two sequential operations using up to three data sets or constants. If you want to do more than 2 math operations, you will have to spread it among multiple pages. The same buckets can be used in the calculations on different pages, but you should use a different results bucket for each page.

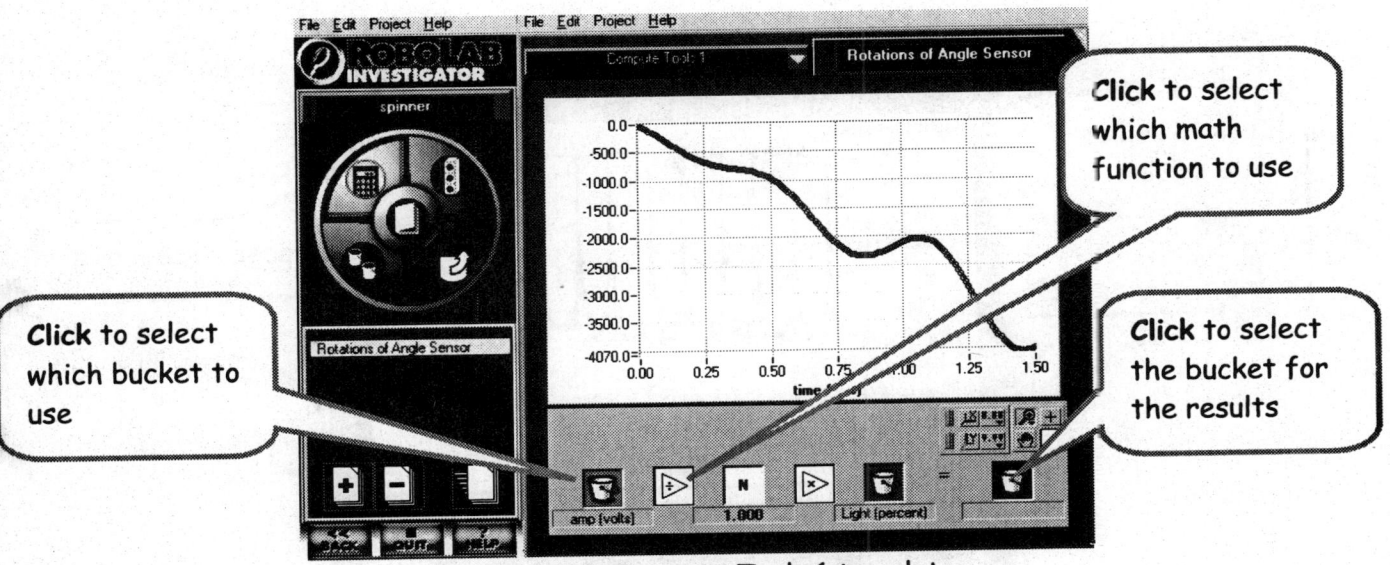

Figure 3.66. Compute Tools 1 template.

All 10 buckets are available for use, along with a numeric (floating point) constant. There are 9 different algebraic operations available: add, subtract, multiply, divide, sine, cosine, tangent, exponential, and natural logarithm.

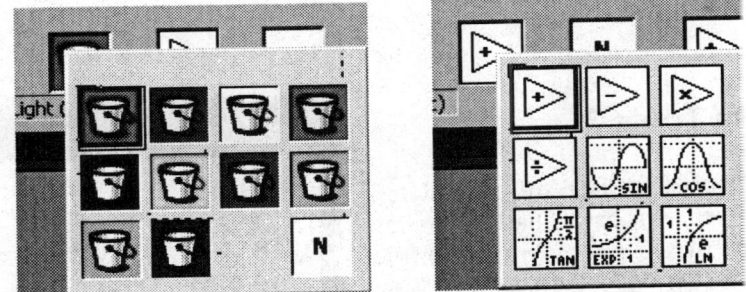

Figure 3.67. Compute Tools 1 Bucket and Math pop-up menus.

Compute Tools 2

Compute Tools 2 is used to plot one data set against another. Unlike **View and Compare Area**, here we can select which part of the each data set we want to use. Since each bucket actually contains 3 columns of data (data point number (**N**), time (**X**), and value (**Y**)), we have a choice of three values to choose from. In the figure below we've plotted **N** versus **Y** instead of the standard **X** versus **Y** (take a close look at the axes labels).

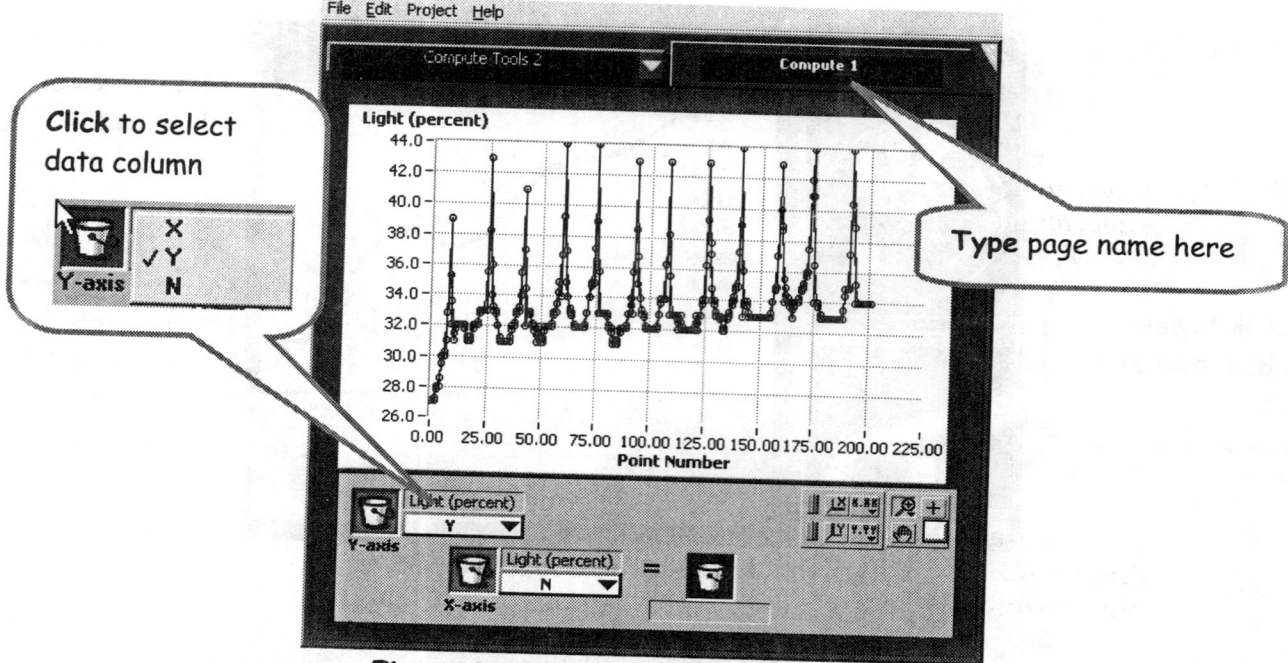

Figure 3.68. Compute Tools 2 template.

Compute Tools 3

Compute Tools 3 is used to manipulate a single data set with basic math functions. Two sequential operations can be performed on any of the 10 data buckets. As always, be sure to use a different bucket for your results for each Compute Tools page.

In figure 3.69, we've differentiated the light sensor data in the red bucket and stored this processed data in the light blue bucket.

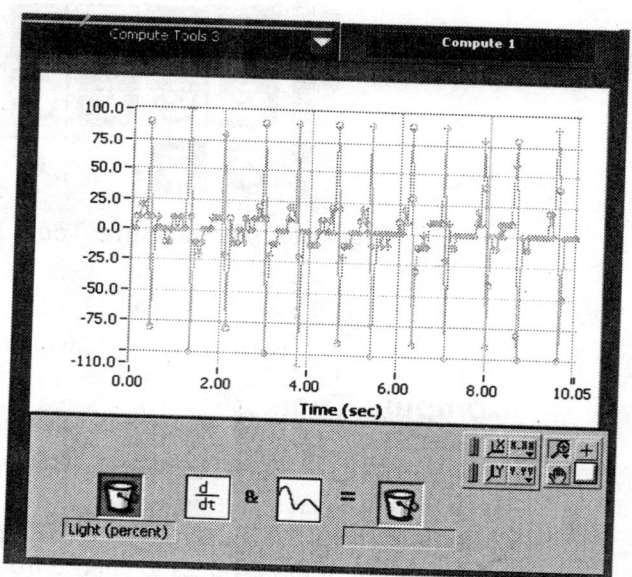

Figure 3.69. Compute Tools 3 template.

There are 11 different math functions available in Compute Tools 3: no change, minimum, maximum, mean, standard deviation, slope, area under the curve, derivative, integral, average lines, and linear curve fit.

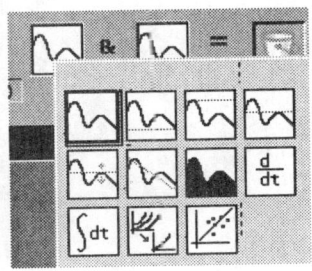

Figure 3.70. Compute Tools 3 functions pop-up menu.

Interesting from a numerical methods standpoint, integrating the light blue bucket does not result in the recovery of the original data, which was in the red bucket (compare Figures 3.68 and 3.71).

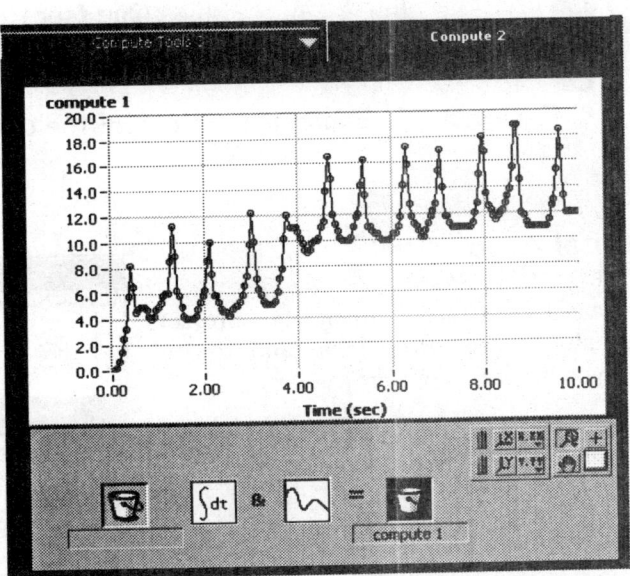

Figure 3.71. Integrating the differentiated data does not recover the original data set!

Exporting Data

After you've manipulated the data, it can be exported to a text file as shown in Figure 3.72.

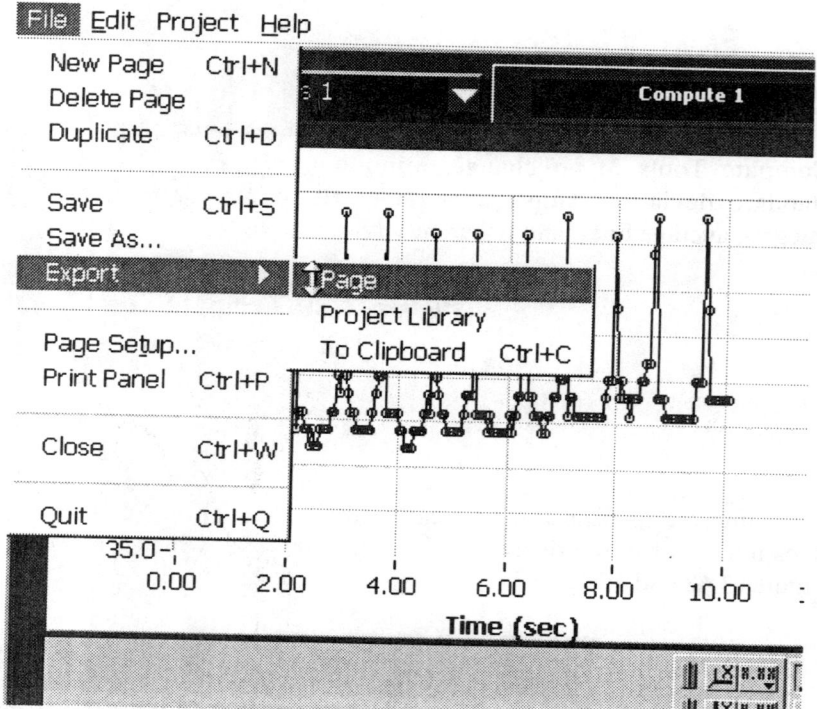

Figure 3.72. Exporting data from the Compute Area.

3.6.5 Journal Area

The journal area is used to document your project. You can include graphs, programs, and digital images to help you describe your project.

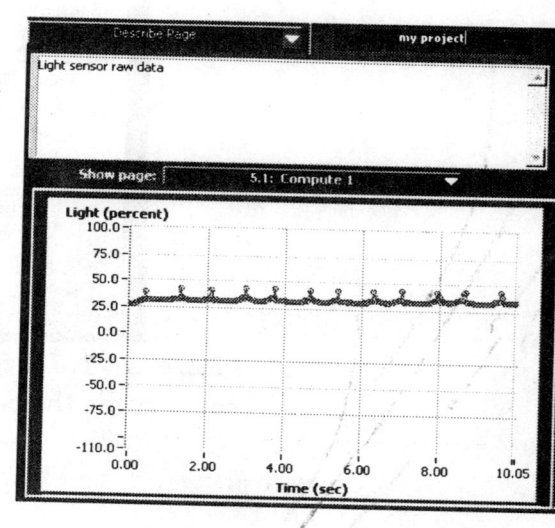

Figure 3.73. Journal Area

CHAPTER 4
BLACK LEVEL

Skill badges available in this Chapter

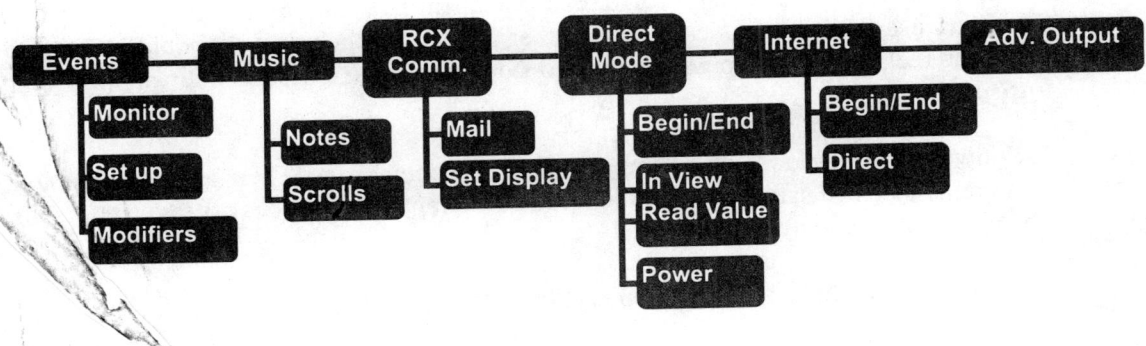

4.1 Black Challenges

All Black Level Design Challenges presume that you have already earned all the Green and White Level skill badges.

4.1.1 LEGO Slot Machine

Challenge: We're from Nevada so the one-armed-bandit was bound to make an appearance in this book sooner or later. The objective of this challenge is to make a LEGO slot machine which randomly picks three numbers between 1 and 4 and if all three numbers match, do something spectacular.

Skill Badges: Events

Procedures:

Experimental Setup: Nothing special needed, except a gaming permit if any money changes hands.

Robot Design: Rather than use reels like actual slot machines, for this challenge we will use the LCD to display the numbers. Each random number selected will be displayed using one of the digits on the LCD. For example, if the numbers were 1, 4 and 2, then 142 would be displayed on the LCD. The user should see the numbers on the LCD changing randomly to simulate the spinning of the reels.

Your version of a slot machine should use a light or touch sensor to begin the "spinning" of the reels and then stop one number every 2 seconds (for a total of 6 seconds). Thus, after the spinning starts all three digits should be randomly changing. After 2 seconds, 1 digit is fixed and the other 2 are still randomly changing. After 4 seconds, 2 digits are fixed and the third is randomly changing. Finally, after 6 seconds all three digits should be fixed and your program should check for a winner (such as 333) and do something creative if a winner is detected.

Program: To program the slot machine you will need to use **containers**, **events** and the **set display** function. You actually could do this challenge without events, but we will insist that you use them for this challenge. Either a light or touch sensor should be used to start the reels spinning. To get the reels to stop spinning at 2 second intervals, <u>you must use an **timer event**</u>.

A flowchart for a sample algorithm is shown on the next page. There are many possible programs that will satisfy this Challenge. The example shown does not use a task split, but your solution may.

Hints: Each task can have its own independent event start and land.

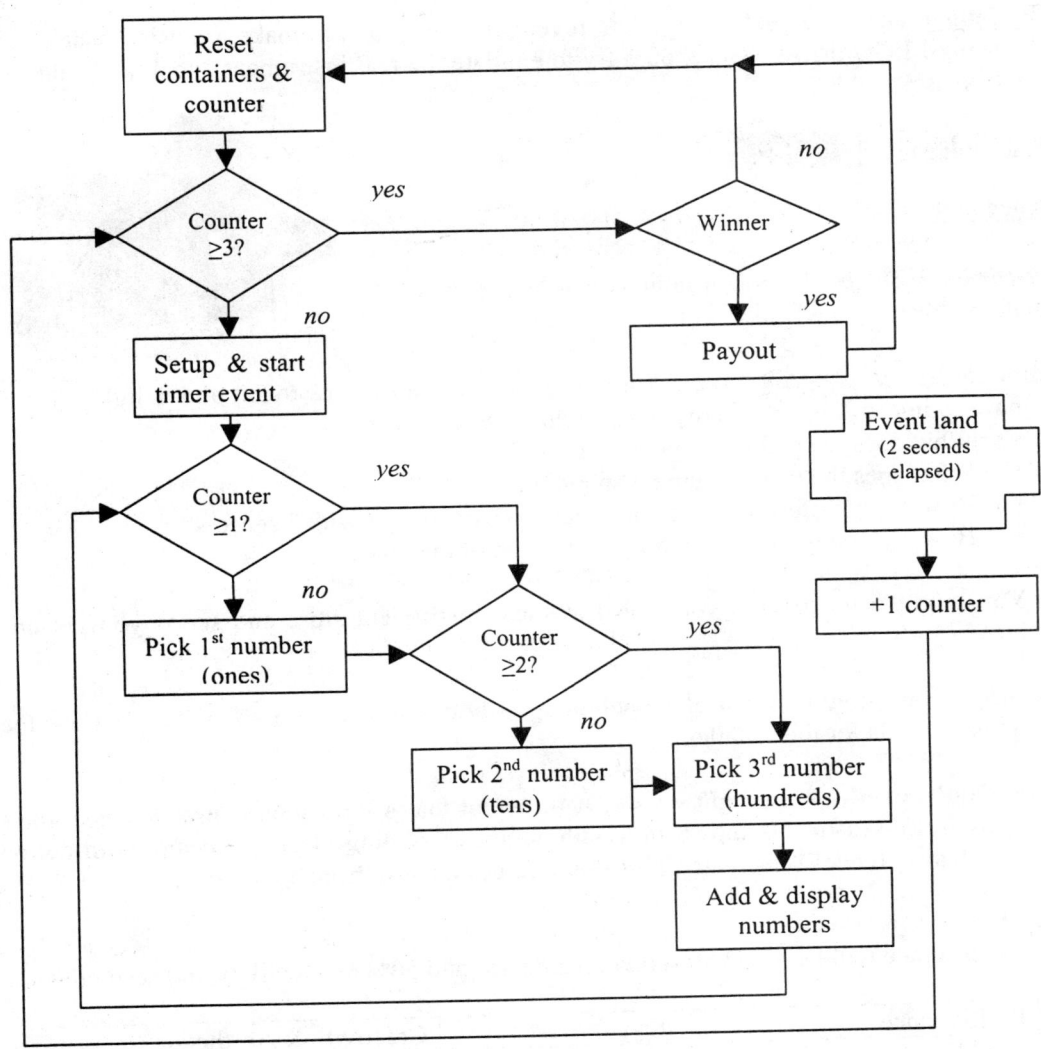

Grading:

Your grade will be based 50% on performance and 50% on creativity and aesthetics.

Performance	Creativity & Aesthetics
A: Your slot machine works perfectly	A+: Best of show
B: "Spins" and stops, but doesn't pay winners	A: Outstanding
C: The reels "spin" but don't stop	B: Good
D: The reels don't "spin"	C: Okay
F: You don't show up to class	D: Nothing special
	F: Divert your eyes!

4.1.2 The Sound of Music

Challenge: The band geeks finally get revenge! Your job is to make a musical instrument using LEGO bricks. You should try to emulate the real instrument's look and sound as best as possible.

Skill Badges: Music

Procedures:

Experimental Setup: No special setup is required for this Challenge.

Robot Design: Since you have a limited number of sensors (2 touch and 1 light), you'll have to improvise a bit. Here are some things to consider:
- What does the real instrument look like?
- What does the real instrument sound like?
- How many different sound combinations can you get with 3 sensors?
- How are you going to control the duration of the notes?

You might want to visit your local musical instrument store and see if you get any ideas.

Program: Your program must play each note individually. You are not allowed to use the music scrolls for this Challenge.

Hints: Don't' overlook the light sensor. It turns out that a light sensor also makes a good proximity sensor. As shown in the photo above, sliding the white beam in an out of the black "tunnel" results in light sensor values ranging from 30 to 60.

Grading:

Your grade will be based 50% on performance and 50% on creativity and aesthetics.

Performance	Creativity & Aesthetics
A+: You play a duet with another team	A+: Looks like the real thing
A: We can recognize the song	A: Outstanding
B: Looks like an instrument & plays sounds	B: Good
C: Makes noises	C: Okay
D: Looks like an instrument	D: Nothing special
F: You don't show up to class	F: Divert your eyes!

4.1.3 Bionic Bat

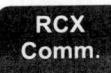

Challenge: The objective of this challenge is to make a proximity detector using the light sensor and the RCX infrared communications port. You detector should play ascending notes as it gets closer to an object.

Skill Badges: [RCX Comm.] AND [Music]

Procedures:

Experimental Setup: Nothing special is needed for this challenge. However, a fairly dark room with no external windows (which can act as significant IR sources) is helpful.

Robot Design: The physical appearance of the robot will count towards creativity. However, in terms of functionality on the RCX and a light sensor are required (as shown in the photo). See section 4.4.3 to get started on investigating the behavior of an IR proximity detector.

Program: The **send mail** function will be used to activate the IR emitter on the RCX. Since the IR signal is not continuous (it's actually a series of short pulses), you will have to develop a way of filtering the light sensor readings. You might consider either averaging the readings or using the maximum value during a short time interval.

Your "bat" should emit a tone at all times (to help it navigate of course). The pitch should be inversely proportional to the distance from the object. That is, as you get closer, the pitch should increase.

Hints: You should make sure the IR power on the RCX is set to high. This is done via the Administrator tab on the introductory ROBOLAB window. Also, the maximum frequency at which you can repeatedly send mail will depend on the total size of your program, so try to be conservative with your code (especially task splits).

Grading:
Your grade will be based 75% on performance and 25% on creativity and aesthetics.

Performance	Creativity & Aesthetics
A: You can navigate with your eyes closed	A+: Best of show
B: Your bat works occasionally bumps into walls	A: Outstanding
	B: Good
C: Your bat detects walls, but not reliably	C: Okay
D: Your bat makes noise	D: Nothing special
F: You don't show up to class	F: Divert your eyes!

4.1.4 Speed Walking (revisited)

Challenge: This is an advanced version of the Challenge described in section 2.1.7. Design and build the fastest walking robot you can. Just to make things interesting, you will be racing over a pebble surface. The "start gun" will be in the form of RCX mail.

Skill Badges:

Procedures:

Experimental Setup: This challenge is typically run over a pebble course in a single elimination format with the winner of each race advancing to the next round. The instructor will also need to have an RCX ready to send out the mail values (be sure to set IR power to high). The program should send out several "false" signals to keep things interesting.

Robot Design: For this exercise you need to design and construct a robot that can walk (crawling may also be permitted) over a mildly uneven surface.

Most running races, begin with the famous phrase "On your mark, get set, go!" In this challenge, we will simulate this phrase with a series of 3 mail values: 11, 12, and 13. Thus, your robot should react to each mail value, but not begin racing until the last number (13) is received. For the first 2 mail values (11 and 12) you should think of something creative to do without starting the race (i.e. don't move forward).

The only restriction is that you cannot intentionally trip another robot.

Program: You must program your robot to react to the 3 different mail values (11, 12 and 13). Your robot should not react to any other mail values or repeated mail values.

Grading:

Your grade will be based 25% on performance and 75% on creativity and aesthetics. Creativity will be based on your "on your mark" and "get set" actions in addition to the walking/crawling mechanism employed.

Performance	Creativity & Aesthetics
A+: You are the fastest walker	A+: Best of show
A: You walk, react, and win at least once	A: Outstanding
B: You walk and react to mail correctly	B: Good
C: You walk, but react to the wrong mail	C: Okay
D: You walk, but don't react to mail	D: Nothing special
F: You don't show up to class	F: Divert your eyes!

4.1.5 Simon Says

Challenge: You know the rules of the game, now we want you to make a robot that plays it as well as you do. Your instructor will build a robot that acts as "Simon," and your robot has to do is exactly what we say, or suffer the consequences. The last robot standing wins in this battle of wits and deception.

Skill Badges: `RCX Comm.`

Procedures:

Experimental Setup: The instructor (or TA) will have an RCX programmed to send out mail values ranging from 0-255 (it should send the value 4 times in a row in rapid succession). The RCX should be programmed to display the mail value sent to the LCD panel so everyone knows what value was just sent. Typically, up to 10 student robots can participate simultaneously.

Robot Design: Since the RCX can't decipher voice commands (yet) Simon will be "talking" to your robots over the IR port. You can read and write to the IR port by using the Mail commands in ROBOLAB.

Everybody lines up their robots with the IR transmitter/receiver facing Simon. Your robot must perform a specific action based on the number Simon sends to you via the IR port. We will send the number four times consecutively so nobody misses the signal. For this reason, do not check the IR port while you are performing the requested action (i.e. don't do the action 4 times in a row). Simon will wait long enough for everybody to complete the action before sending out another signal.

Program: If Simon sends out a:
1. Your robot must turn to the right 90 degrees (approximately) and turn back to face Simon.
2. Your robot must turn left 90 degrees and turn back to face Simon
3. Your robot must drive forward for 6 inches, play a tune for 5 seconds, then return to its original spot.
4. Your robot must drive back 6 inches, do a full 360 and drive forward to its original spot

If Simon sends any other number, you should do nothing.

The Goal - don't get knocked out: if your robot performs the wrong task, or moves when it shouldn't, it's going to be obvious, since it will stand out from the competition like a sore thumb.

Hints: Try to make your motions as precise as possible, since having your IR port in view of Simon's is essential to lasting the whole round. In order to make two motors behave the same (to drive straight forward without curving, for instance), you may need to run them at different power levels. Experiment.

Grading:

Your grade will be based 75% on performance and 25% on creativy and aesthetics.

Performance	Creativity & Aesthetics
A+: You're the last one standing	A+: Best of show
A: Your robot performs admirably, performing all the correct functions	A: Outstanding
B: Your robot gets dumped early on due to very imprecise movements or incorrect actions	B: Good
	C: Okay
C: Your robot does the wrong move the first round	D: Nothing special
	F: Divert your eyes!
D: Your robot is left at the line, wondering what happened	
F: You don't show up to class	

4.1.6 Demolition Derby

Challenge: This Challenge requires cooperation between two teams. The objective is to build two LEGO cars that race straight towards each other and then stop, narrowly avoiding a head-on collision.

Skill Badges:

Procedures:

Experimental Setup: All that is required is a flat surface with two lines 4 feet apart that indicate the start location of each vehicle.

Robot Design: Your two LEGO vehicles will start facing each other 4 feet apart. You accomplish this daredevil feat by constantly communicating, via the IR port, the distance your car has traveled to your partner. By knowing how far you and your partner have traveled, each of you should be able to calculate the remaining distance between your vehicles and determine when to stop.

One obvious solution is for only one of the cars to actually move. Which is okay, except that you will get a D grade if you take this approach (both teams get the same grade in this Challenge). Neither car can stop moving for more than 2 seconds if you want to earn a C or better grade.

Finally, to stress the point that engineering is a conservative profession, if you're cars crash into each other (or pass each other if your aim is poor), then you will get a C. It is much better to stop way short than to go too far.

Program: The IR port cannot send and receive data simultaneously, so it won't help if you are both sending data at the same time. Thus, you will have to work out a *handshaking* protocol that defines when you will send and when you will receive data. To measure how far you've gone, you can either go by time or build an encoder (see chapter 3.1.8).

Hints: You many want to set the IR power level to high (see chapter 1.3.1)

Grading:

Your grade will be based 75% on performance and 25% on creativity and aesthetics (more creativity points will be given if you cars are very different).

Performance	Creativity & Aesthetics
A+: Fastest pair to within 9 inches	A+: Best of show
A: You stop within 9 inches of each other	A: Outstanding
B: You run, communicate, and stop without crashing into each other.	B: Good
	C: Okay
C: You crash into each other - be conservative!	D: Nothing special
D: One of the cars stops for more than 2 seconds	F: Divert your eyes!
F: You don't show up to class	

4.1.7 Animal Behavior

Challenge: This is civilized head-to-head sumo wrestling battle. Just like real sumo wrestling, the objective is to push your opponent out of the ring. However, your robots must display some *animal behavior* at the start and end of the battle to make things interesting.

Skill Badges: [Events] AND [RCX Comm.]

Procedures:

Experimental Setup: The competition arena is a white circular area 3 feet in diameter. The outer edge of the competition arena is lined with silver (mirror-like) tape. Approximately 6 inches inward is a concentric ring colored black. The competition can be autonomous or tethered (the instructor will inform you). Your instructor may also place a limit on the size of your robots.

The instructor will also have an RCX or two on hand that can send and receive the 3 mail commands just in case things go awry. Typically, most problems occur to trying to receive the surrender mail signal.

Robot Design: In the Animal Kingdom, animals fight all the time over territory, mating rights, food, etc. A typical encounter (typically between two males) starts with *posturing* (grunting, beating chest, or stomping feet). Then the actual battle commences. However, animals rarely fight to the death. One animal, the loser, *submits* (runs away, puts it's tail between it's legs, or rolls over belly up). The winner then often performs some kind of *victory* behavior just to rub salt into the wounds.

In this competition your robot will have to be able perform three kinds of behaviors:
Posturing – try to intimidate your opponent.
Submission – "I give up!"
Victory- "Yeah, and don't come 'round here anymore!"

The rules of the game are simple, start by taking turns performing your posturing behaviors. If you get pushed out of the competition ring (i.e. reach the outer silver tape) you lose and must submit to your opponent. If you win, you must perform a victory behavior.

Program: Since you won't know until the battle whether you go first or second, you should have 2 programs that are nearly identical. If you go first, you should perform your posturing behavior and then send mail (value = 10) to your opponent. You should then wait until you receive a mail value of 20 back from your opponent before commencing to seek and destroy.

If you go second, you should wait patiently until you receive mail with a value of 10. Once this happens, it's your turn to display your posturing behavior. When you are done showing off, you should send mail with a value of 20 and begin fighting!

Both robots **must use an Event** to monitor the outer edge of the ring (the silver tape). If you cross the outer edge, you robot should perform its submission behavior and start continuously sending out mail with a value of 30.

If you detect mail with value of 30 at anytime, this is a clear indication that your opponent has given up (submits to your overwhelming dominance) and your robot should perform its victory behavior.

Your robot should not react to any mail values other than 10, 20 and 30. The instructor may have an RCX on hand that will send out "false signals" just to be diabolical.

Hints: Be creative with your behaviors. Your light sensor should point downwards to detect when you are getting close to the edge (the black line) and when you've lost (the silver line). If you reach the black line, you may want to perform some kind of "evasive" maneuver. Finally, don't block the IR port or you won't be able to send/receive mail.

It is also important to end up facing your opponent after posturing or you won't be able to send or receive mail effectively.

Grading:

Your grade will be based 30% on performance and 70% on creativity and aesthetics. The 3 behaviors will obviously play an important role in evaluating the creativity of your design.

Performance	Creativity & Aesthetics
A+: Undefeated champion	A+: Best of show
A: You win more than once	A: Outstanding
B: You battle and display the posturing and submit/victory behaviors on cue	B: Good
	C: Okay
C: You posture and start battling on cue	D: Nothing special
D: You have something that moves	F: Divert your eyes!
F: You don't show up to class	

4.1.8 Can you hear me now?

Challenge: The objective is to send a short coded message over the Internet.

Skill Badges: Internet

Procedures:

Experimental Setup: Two computers with Internet access are required for this challenge. Ideally, the computers are located in different rooms. The instructor will configure the ROBOLAB Internet Server software on the host computer.

Each team must have at least one person to program and one person to receive the message. During class, the instructor will specify to the programmers a short message (3 or 4 letters). The programmers must then write a ROBOLAB program and send it via the internet to their partners, who must execute the program on their robot and decode the message.

Robot Design: Your team will have to determine how to decode a message. There are lots of ways: Morse code, flashing lights, alphabet wheel, LCD display, Braille, etc. Creativity will be largely based on the method you choose to decode the message.

Program: The program will depend on the method your team selected to decode messages. You can practice without using the Internet, but be sure you know how to send and receive using ROBOLAB Internet Server before class.

Hints: You may want to include some kind of error checking scheme in your program.

Grading:

Your grade will be based 50% on performance and 50% on creativity and aesthetics.

Performance	Creativity & Aesthetics
A: You successfully decode the message	A+: Best of show
B: Your program is received and runs	A: Outstanding
C: Your send the program somewhere	B: Good
D: You show up to class with something	C: Okay
F: You don't show up to class	D: Nothing special
	F: Divert your eyes!

4.1.9 CodeMaster

Challenge: The objective is to decode a short message based on container values. The trick is that the container values will be set using **direct mode** while your decoder program is running.

Skill Badges: [Direct Mode]

Procedures:

Experimental Setup: computers with ROBOLAB and IR towers are required for this challenge.

Robot Design: In his book, *Creative Projects With LEGO Mindstorms*, Ben Erwin describes the CodeMaster developed for sending secret messages. This is a similar design task in that you must somehow translate a container value into letter. There are lots of ways of communicating text: Morse code, flashing lights, alphabet wheel (see next page), LCD display, Braille, etc. Creativity will be largely based on the method you choose to decode the message. Displaying a number to the LCD and then looking up the corresponding letter will not earn you very many creativity points.

Program: The exact program will depend on the method your team selected to decode messages. However, all programs must use containers User3 through User6 to store the 4 numbers. Each number will correspond to the letter's place in the alphabet. That is, 1=A, 2=B, and so on through 26=Z. Once the first non-zero container value is detected by your decoder program, it should start decoding the 4 letter message.

During class, the instructor will tell you the message that must be decoded by your robot. With your decoder program already running, you must use **direct mode** and write a program that fills each of the containers (User3-User6) with the appropriate number.

Hints: Typically, the faster your decoder moves (if it moves at all), the more prone to error it will be.

Grading:

Your grade will be based 75% on performance and 25% on creativity and aesthetics.

Performance	Creativity & Aesthetics
A: You successfully decode the message	A+: Best of show
B: Message received but decoded wrong	A: Outstanding
C: Direct mode program works	B: Good
D: Your decoder program runs	C: Okay
F: You don't show up to class	D: Nothing special
	F: Divert your eyes!

4.1.10 Reverse Engineering

Challenge: The objective of this challenge is to reverse engineer another team's program that is sent to you over the Internet.

Skill Badges: Internet

Procedures:

Experimental Setup: Two computers with Internet access are required for this challenge. The computers do not need to be located in different rooms. The instructor will configure the ROBOLAB Internet Server software on the host computer.

Robot Design: Your team should build a robot and write a program to download to it over the Internet. In class, you should give your robot to another team and then send your program over the Internet. The other team (the one with your robot) will then play with your robot and try to determine what your program looked like. Your robot must utilize at least 2 sensors and one motor.

While they are busy trying to reverse engineering your code, you will be busy reverse engineering another team's code.

Program: Unless you derive pleasure from watching your classmates sweat, you should create a program with only moderate complexity. To make things fair, your program cannot contain any mail functions and must contain at least one fork and one loop/jump.

Hints: In order to reverse engineer code, you will probably have to run the program and test the reaction to the sensors many times.

Grading:

You should submit both your original program and the other team's reverse engineered program. Be sure to clearly label which team reversed engineered your program. Your grade will be based 70% on how well you reverse engineered the other team's program and 30% on how well your program was reverse engineered by them.

Performance
A: You essentially reverse engineer the entire code
B: You get most of the main program features
C: You reverse engineer some of the code
D: You get a program over the Internet
F: You don't show up to class

4.1.11 Stay Away From the Light

Challenge: The objective of this Challenge is to investigate the basics of feedback control by building a robot that adjusts the motor power level according the light level.

Skill Badges: Adv. Output

Procedures:

Experimental Setup: All that is required is a flat surface and a flashlight.

Robot Design: This project is an advanced version of the light following project (see 3.1.5). Instead of simply heading for the light, this time you need to build a robot that will home in on a specific light intensity (specified by the instructor). When the intensity is too low, the robot should drive forward. When the intensity is too high, the robot should back up. Thus, the robot should go towards the light, but not get too close.

The **error** is defined the difference between the desired and actual light intensities. In this challenge, you need to set the motor power to be proportional to the error (thus the term, ***proportional control***). If the error is positive (desired intensity is greater than actual intensity) the robot should back up. The larger the error, the faster (more power) is backs up. If the error is negative, the robot should move forward (towards the light), again with the motor power proportional to the magnitude of the error.

In terms of robot construction, any LEGO car will do. The only caveat is that the car should not move at power level 1. That is, at power level 1, the motors should stall. Play with various gear ratios and wheel diameters to find a combination that works. We also recommend using an AC adapter, as stalling a motor will drain the batteries very quickly.

Program: You will use Proportional control (a.k.a. P-control) to make the robot move towards the light, but not get too close. At least three containers are needed (you may use more depending on your approach): set point (desired intensity), P-gain, and motor power (an integer from 0 to 7). A sample 2-task algorithm flowchart is shown on the next page.

Try lots of different P-gain settings to investigate the different responses. In general the lower the P-gain, the longer it will take for the robot to "settle down". But increasing the P-gain can cause some strange behaviors! Can you get both stable and unstable behaviors using the same program but different P-gains? In-class you must demonstrate at least 3 different P-gain settings.

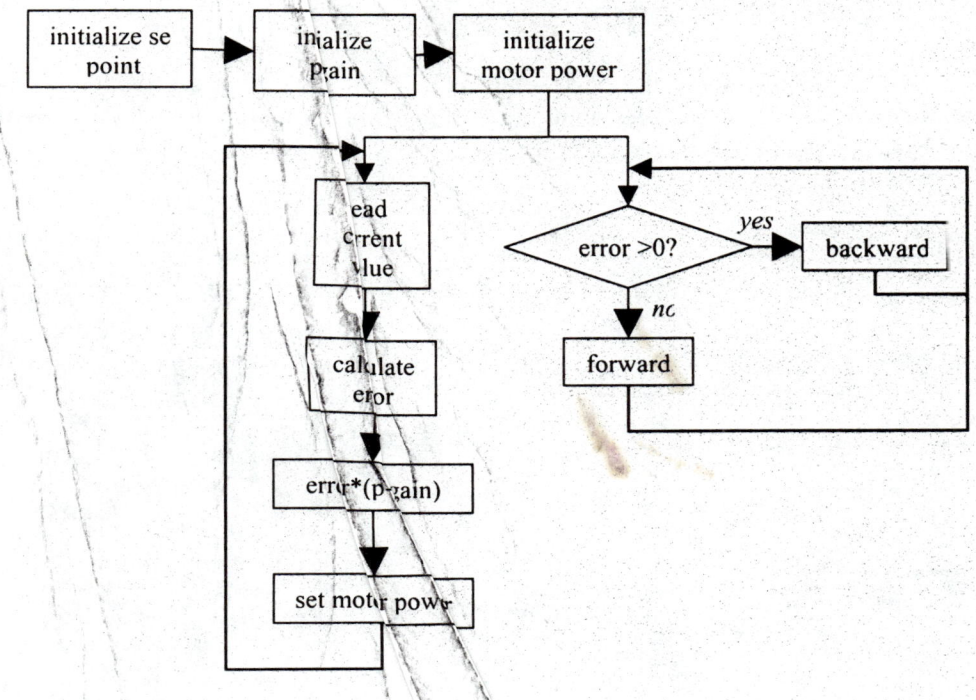

Grading:

Your grade will be based 80% on performance and 20% on creativity and aesthetics (more creativity points will be given if your cars are very different).

Performance	Creativity & Aesthetics
A: You demonstrate 3 different responses	A+: Best of show
B: You demonstrate 2 different responses	A: Outstanding
C: You demonstrate 1 response	B: Good
D: One show up to class with a robot	C: Okay
F: You don't show up to class	D: Nothing special
	F: Divert your eyes!

4.2 The Events Badge `Events`

The Events badge covers the basic concepts of an event: what an event is, how you program an event, and the rules about multiple events. The more advanced event concepts (e.g. hysteresis) will be covered in Chapter 6.

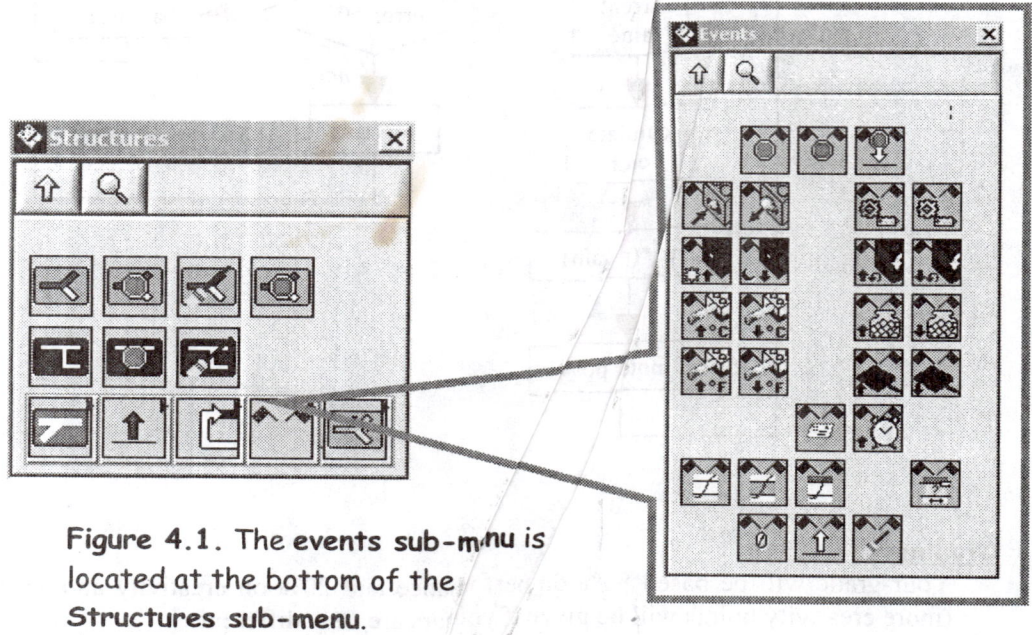

Figure 4.1. The **events sub-menu** is located at the bottom of the **Structures sub-menu**.

4.2.1 What's an event?

An *event* is like a special combination of a *wait for* and a *jump* that runs in the background. When the specified *event* occurs, the program jumps to the *event landing* location.

The program below is an example of using an *event* to jump out of an infinite loop created with a *jump/land* pair. The program sets up a *red event* that is triggered when touch sensor 1 is pressed, starts monitoring for the *red event* and then enters an infinite loop that turns on *Motor A*. When the *red event* occurs (i.e. touch sensor 1 is pressed), the program jumps out of the infinite loop and lands at the *event landing* and then stops *Motor A* before ending.

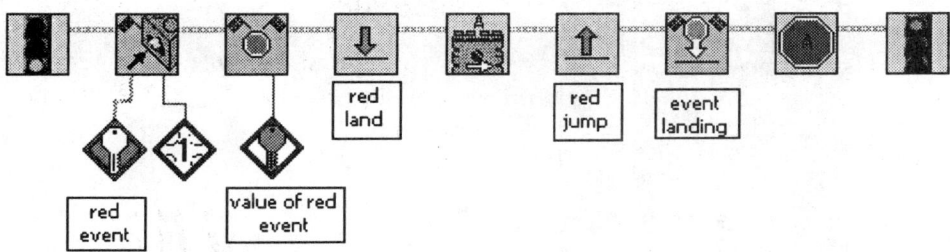

Figure 4.2. The red event is used to jump out of the infinite loop.

4.2.2 How to program an event

All events follow the same basic format:

1. *Set up the event* – define what condition you want to wait for
2. *Start monitoring event* – tell the RCX to start waiting for the event(s)
3. *Event landing* – this specifies the location in the program that ALL the events will jump to.

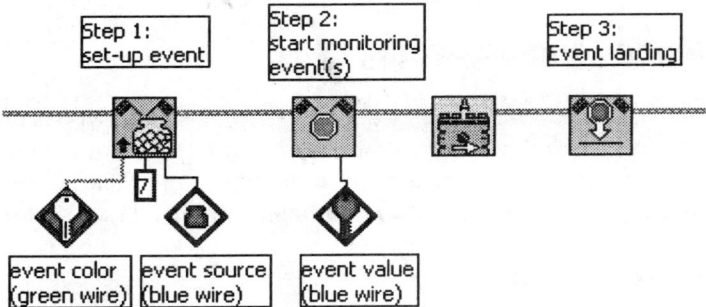

Figure 4.3. The 3 basic steps in using an event and the 3 basic event modifiers required to setup and monitor an event.

 Step 1: Setting up an Event

Setting up an *event* is very similar to using the familiar **wait for** command. Figure 4.4 shows the most common types of events that we will use (for now, we'll ignore the generic event). We can set 16 types of *events* ranging from a **touch event** to a **light event** to a **timer event**.

Page 4-20

Figure 4.4. The 16 common types of events to set up.

In addition to defining what you want to monitor (Figure 4.4), you also have to define the *event source*, and which *event* (red, blue, or yellow) you are setting up. The *event source* is *sensor port value, container value, mail value,* or *timer value,* depending on the type of event. There are 3 basic events (red, blue, and yellow) and a generic event in case you want to monitor more than 3 events.

Figure 4.5. Red, blue, and yellow events.

 ### Step 2: Monitoring an event

After you set up the event, nothing actually happens until you start monitoring for the event. This is done using the green *start event monitoring* command and wiring the *value of event* to it (see figure 4.3). If you want to monitor more than 1 event, you should wire all the event values to a single *start event monitoring* command (see section 4.2.4 below)

 ### Step 3: Event Landing

The final step in using events is to define the *event landing* location. The *event landing* defines where ALL the events will land. Even if you've set up 3 different colored events, they will all land at the same location. By definition, the event landing also stops the monitoring of all events. If you wish to continue monitoring events after the landing, you need to restart event monitoring (see below).

 | All events share a single event landing |

4.2.3   Stopping and re-starting event monitoring

We can also stop monitoring and then re-start event monitoring in case you have a part of your program you don't want interrupted for any reason. In the example below, we have stopped the event monitoring during the inner loop and then re-started the event monitoring after the loop. In short, if touch senor 3 is pressed while Motor A is running, the event will be ignored because we temporarily stopped monitoring it.

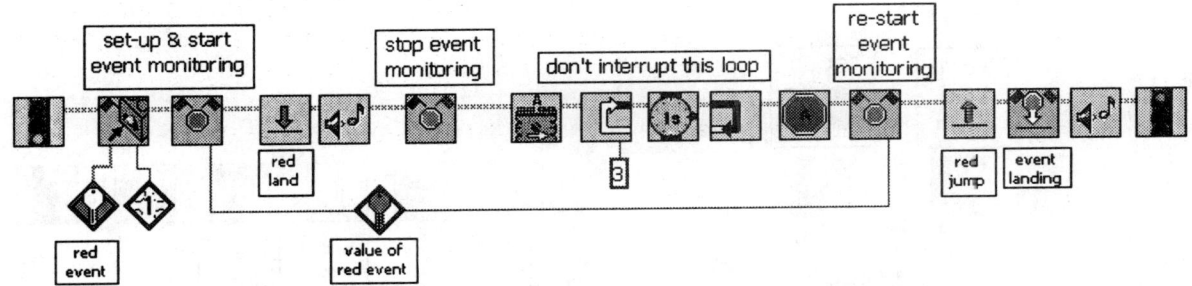

Figure 4.6. Stopping and re-starting event monitoring.

We can also re-start *event monitoring* after the *event landing*, as shown in Figure 4.7. In this example, if the touch sensor is not pressed, the motor will run for a total of 22 seconds. Pressing touch sensor 3 will terminate the loop early and then wait 2 seconds before turning the motor off. However, since we've re-started *event monitoring* after the *event landing*, pressing touch sensor 3 a second time will cause the program to jump backwards (to the event landing). As long as the touch sensor is pressed at least once every 2 seconds, the program will keep jumping backwards from the *wait for 2s* to the *event landing* and the motor will stay on.

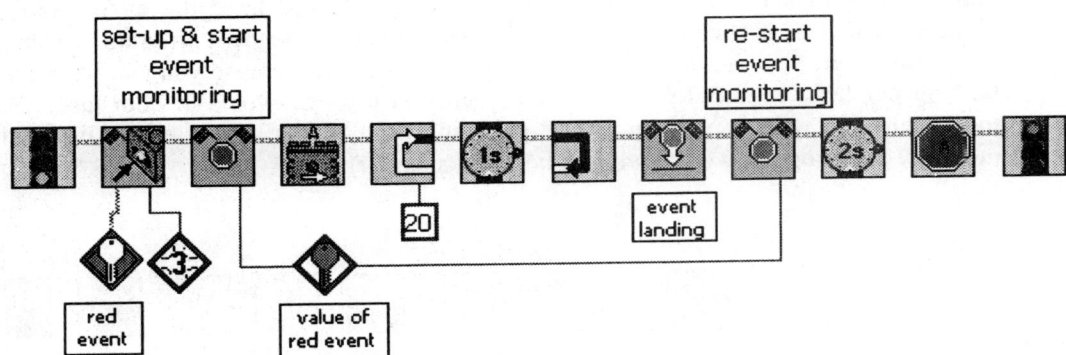

Figure 4.7. Jumping backwards in a program by re-starting event monitoring.

4.2.4 Multiple events

You can monitor up to 16 different events in ROBOLAB, of course you'll have to use the generic event because there aren't enough standard event colors (the red event is event #0, blue is event #1, and yellow is event #2). In the example below, we've setup both the red event and generic event #4. In this case, we can jump out of the infinite loop by either pressing touch sensor 3 or by exposing the light sensor to a bright light. Notice that only one *start monitoring event* command and one *event landing* command are used.

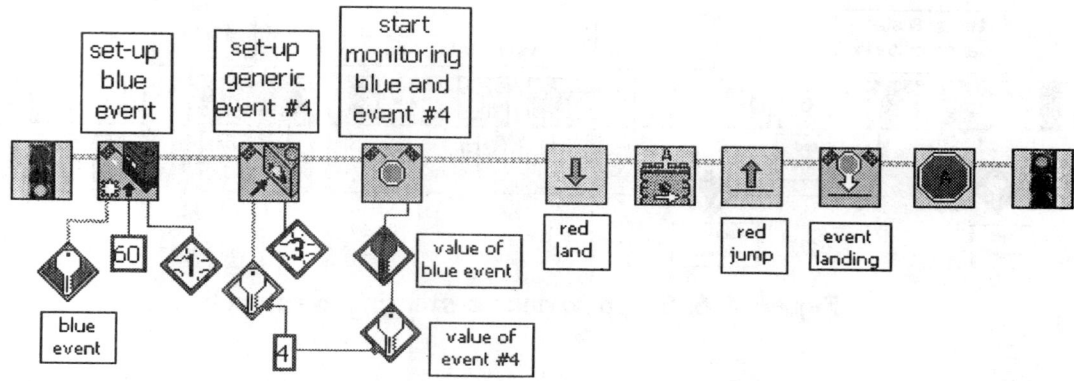

Figure 4.8. You can monitor multiple events, but all events land at the same location.

4.2.5 Events and tasks

It gets a bit complicated when events are used in conjunction with tasks. The basic rule is that you cannot start monitoring an event in one task and have the event landing in a different task. This is because each task is essentially an independent program.

However, you can take advantage this independence because you can have one event landing per task. Figure 4.9 shows a program that utilizes events in both tasks. Pressing touch sensor 3 will stop motor A and exposing light sensor 1 to a bright light will stop motor C. The program won't end until both tasks have ended.

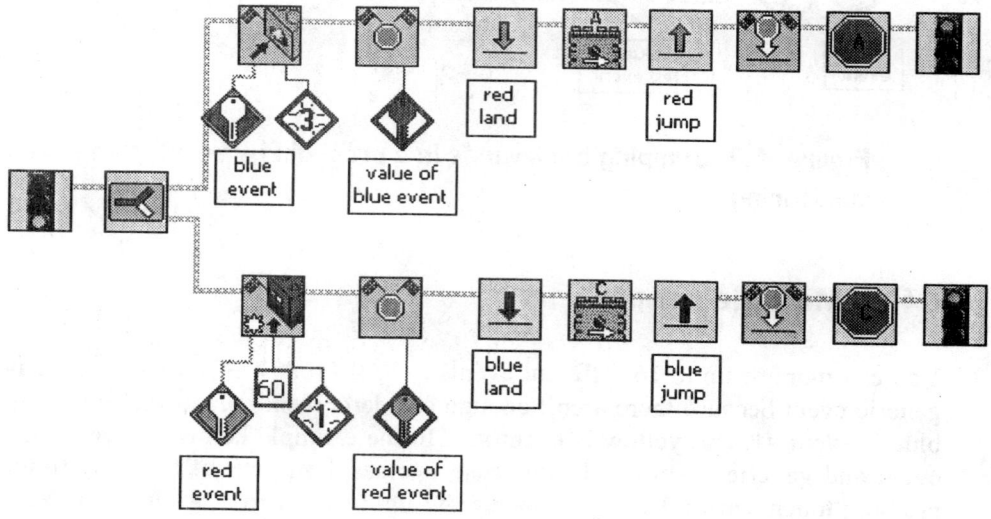

Figure 4.9. Events cannot cross tasks, but you can have one event landing per task.

4.2.6 Event Examples

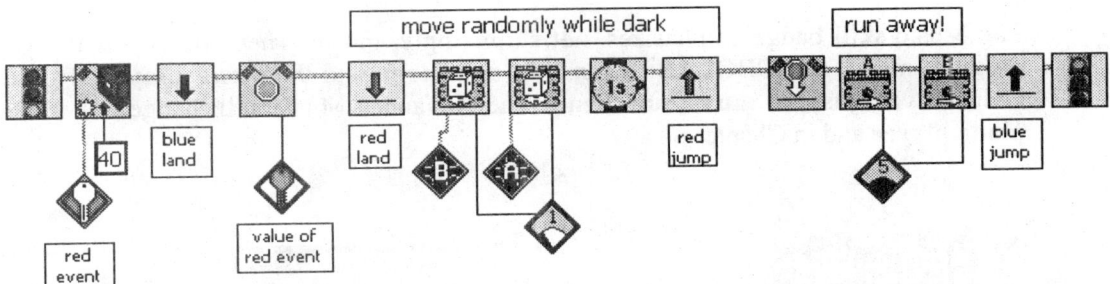

Figure 4.10. An attempt at the LEGO cockroach program done using events – which doesn't quite work correctly!

You may find is strange to give an example that doesn't work as intended, but we find it very educational so here we go. In Figure 4.10 we've tried to make the LEGO cockroach program again, this time using events (see Figures 3.12 and 3.27). Unfortunately, it doesn't quite work correctly. Here's why: the event is actually defined as occurring when the light value transitions from below 40 to above 40 (this is called a *leading edge trigger* in electronics lingo). If the light value is already above 40 when we start monitoring for the event, then, by definition the event has not occurred yet (i.e. it must start below 40 for this to work).

The same will hold true for the other events examples. In Figure 4.9, if the light sensor value starts off greater than 60 at the start, then motor C will not stop until the light value drops below and then rises above 60. Likewise, if the touch sensor is already pressed when the program starts, then it must first be released and then pressed again before Motor A will stop. A very subtle, but important point to understand when using events.

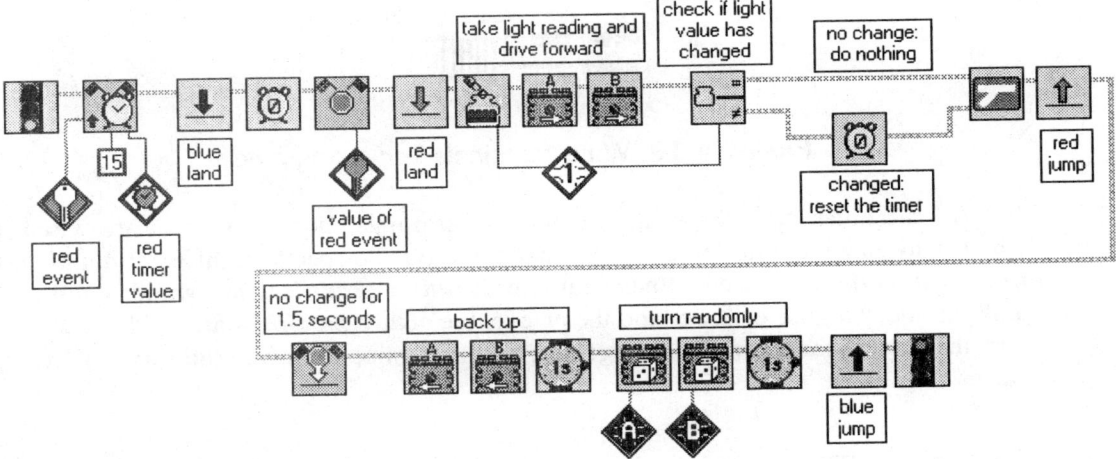

Figure 4.11. Obstacle avoidance using the light sensor and timer event.

Figure 4.11 shows a complex but very effective method of obstacle avoidance using a combination of the light sensor and the timer event. The basic idea is to drive forward as long as the light sensor value is changing. If there is no change for 1.5 seconds, the chances are you have run into something and are not actually moving. If this happens, back up and then turn randomly. A car running this program will randomly roam around the floor, never getting stuck for more than a few seconds. Try it out!

4.3 The Music Badge

The Music skill badge emphasizes, not surprisingly, music. There are two main ways to program music in ROBOLAB: the **music sub-menu** and the **piano player**. This music badge covers just the **music sub-menu**. The Advanced Music skill badge will cover the **piano player** and in Chapter 6.

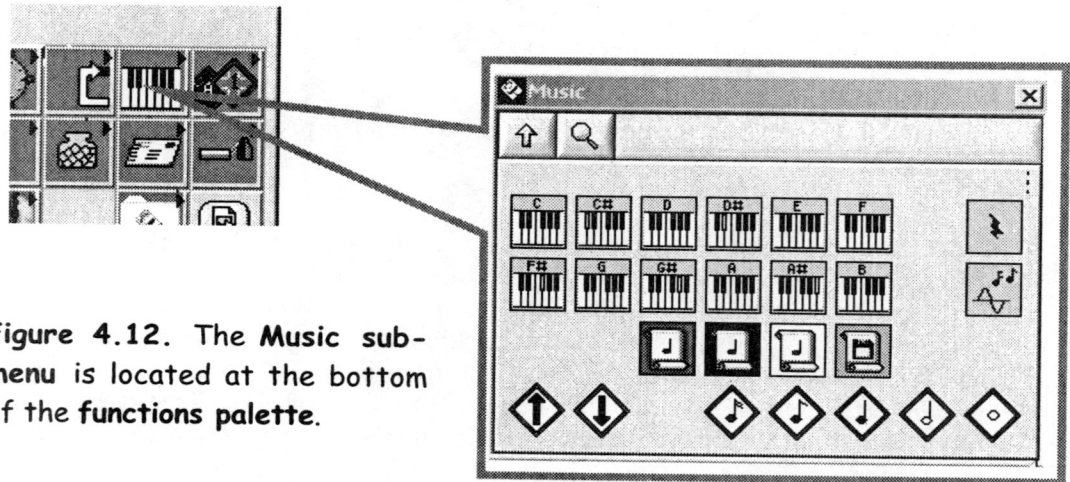

Figure 4.12. The **Music** sub-menu is located at the bottom of the **functions palette**.

4.3.1 Basic Notes

Using music is fairly easy; in fact you may have already used it in one of the White Level Challenges (Chapter 3).

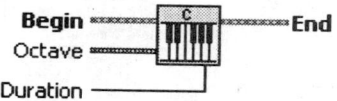

Figure 4.13. Wire terminals for a single note.

All the notes have the same set of wire terminals. If no modifiers are wired, the default is to play a quarter-note on the standard scale. To specify a different duration, the *duration modifiers* are used. When multiple *duration modifiers* are wired together, the durations add. To raise or lower and the octave, the *octave modifiers* are used. By stringing more than one *octave modifier* together, you can raise or lower the note by multiple octaves.

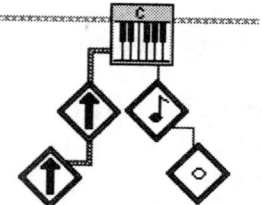

Figure 4.14. Example of raising two octaves and increasing the duration to 1 1/8 counts.

As an example of using both the duration and octave modifiers, we've programmed the first few bars of "On Top of Old Smokey" in the figure below.

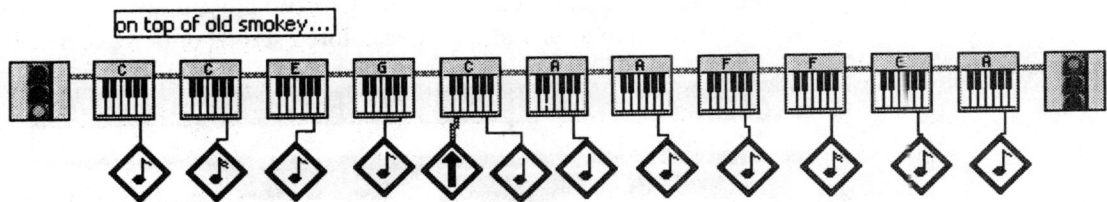

Figure 4.15. First few bars of "On Top of Old Smokey."

4.3.2 Music scrolls

Rather than enter all those notes individually, we could make use of the ROBOLAB Music Scrolls. The program in Figure 4.15 can then be simplified to the one shown in Figure 4.16 below once we have saved the song to the *red music scroll*.

Figure 4.16. Making music with the *red music scroll* function.

By default the red, blue and yellow music scrolls have "Frere Jacques"; "Row, Row Your Boat"; and "Twinkle Little Star" recorded to them. If you want to use any of these three standard songs then you can simply wire them as shown in Figure 4.16.

To get a custom song, like "On Top of Old Smokey," onto the *red music scroll*, we must utilize the **piano player**, which was briefly introduced in Chapter 2. From the **Tools** menu (or projects menu if you are using ROBOLAB version 2.5.0 or earlier), select **Piano Player** as shown in Figure 4.17.

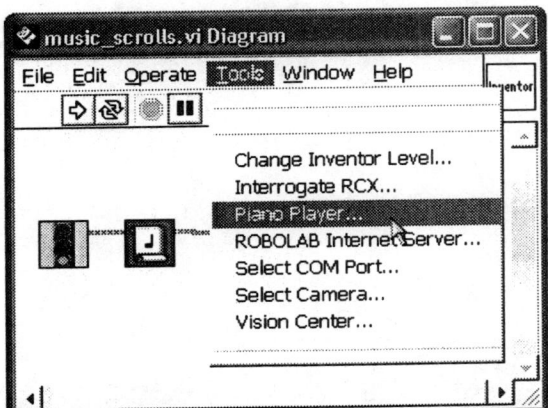

Figure 4.17. The **Piano Player** is accessed via the **Tools** Menu.

The **piano player** actually allows you to do a lot of things, but we're only going to cover the basics here. Refer to Chapter 6 for more information on all the **piano player** functions.

Recording a song to a music scroll is as simple as playing the song on the keyboard and then selecting which music scroll to save it to. If you have the record button on (red = on, gray = off), the notes will appear on the sheet music as you play them. You can select individual notes to change their duration or delete them. When you are satisfied with your creation, you can save your song to one of the three colored scrolls or to a generic file.

Figure 4.18. The **Piano Player** window.

4.4 The RCX Communication Badge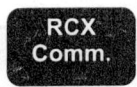

The front of the RCX houses the IR (infrared) communications port. So far you've used the IR port to download programs from your PC to the RCX and to upload data collected from the RCX to your PC. The RCX Communication skill badge focuses on using the IR port to send messages (mail) between two or more RCX's.

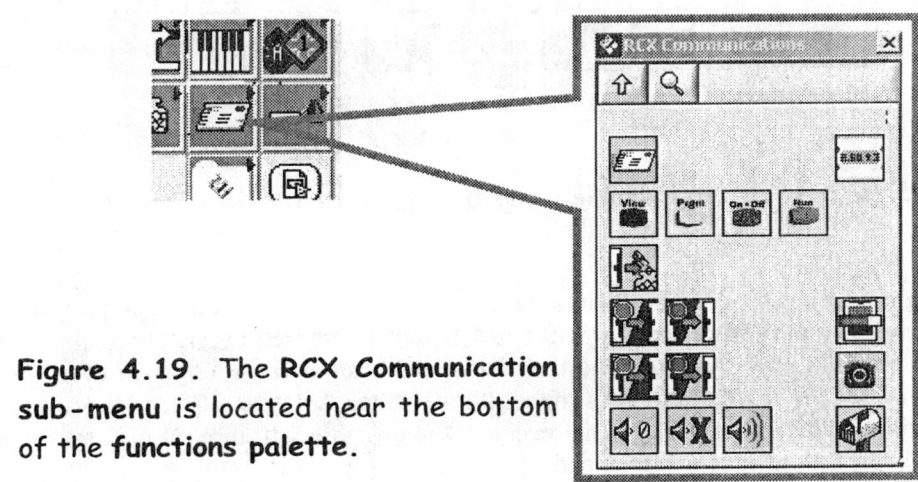

Figure 4.19. The **RCX Communication sub-menu** is located near the bottom of the **functions palette**.

4.4.1 Mail

In ROBOLAB **mail** is any integer between 0 and 255. Mail is sent and received over the IR port. It is important to note that you cannot send and receive mail at the same time. In real life we're all used to the fact the postman will deliver our mail even if we're not home (and it's not a Sunday or holiday). A similar feature exists in LEGO life; <u>you don't have to wait for mail in order to receive it</u>. And, just like junk mail in real life, in LEGO life you don't have the choice to decline mail that another RCX sends – you always accept mail in your mailbox automatically (replacing any existing mail in the process).

To send mail to another RCX you use the *send mail* function. You can also send yourself mail by filling your own mailbox, just in case you can't wait for someone else to send you mail. Both the *send mail* and *fill mailbox* functions are located on the **RCX Communication sub-menu** (Figure 4.19)

If you would rather wait for the mail to be delivered, you can use the *wait for mail* function, which is located on the **wait for sub-menu**. The *wait for mail* function zeros the mailbox and then waits for a non-zero mail value to arrive before proceeding to the next function in the program. It's like sitting at home next to your mailbox waiting for the postman to arrive with an important letter.

Once you've received mail, it's stored in your mailbox. Unfortunately, in ROBOLAB your mailbox is very small and can only hold one piece of mail (i.e. a single integer from 0 to 255). However, there are lots of things you can do with your one piece of mail. There are **mail forks**, **mail loops**, **mail containers**, and **mail events**. You can also use the *value of the mail* anywhere an integer **numerical constant** can be used.

Figure 4.20. Some of the various mail functions.

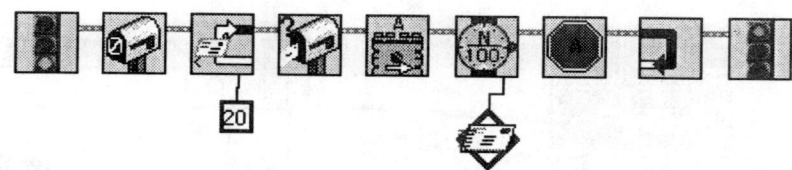

Figure 4.21. An example using the mail wait for, loop, and value.

Figure 4.21 is a program illustrating a few of the mail functions. The program starts by *emptying the mailbox* (sets the mail equal to zero) and then begins looping. Inside the loop, the program *waits for mail* and then turns on **Motor A** for a length of time governed by the *value of the mail*. As long as the mail received is less than 20, the loop will repeat. If mail with a value greater than 20 is received, the program will turn on the motor one last time for that length of time and then end.

Note, in this program the value of the mailbox may change if another RCX sends mail between the *wait for mail* and the *end of loop* functions. To get around this problem, it is best to use a *mail container* rather than the value of the mail. Figure 4.22 shows the same program done with containers.

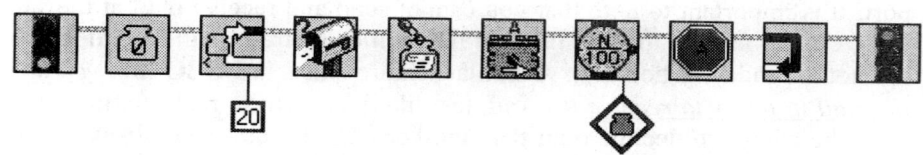

Figure 4.22. Using containers to avoid changing mail values.

Of course, mail is most commonly used for multiple team projects (of which there are many in this book) when two or more robots need to communicate with each other. Since the IR port cannot send and receive data simultaneously, some kind of *handshaking* protocol needs to be worked out so that each robot knows when to "listen" and when to "talk." For example, in Figure 4.23 we've written a program that sends mail ("talks") for 0-1 seconds and then waits for mail ("listens") for up to 1 second. If mail is received, it plays a sound, if no mail is received by the end of 1 second, it jumps back to the start and "talks" again. Try it out with 2 RCX's and you should hear them beeping away at each other, indicating they are holding a conversation successfully.

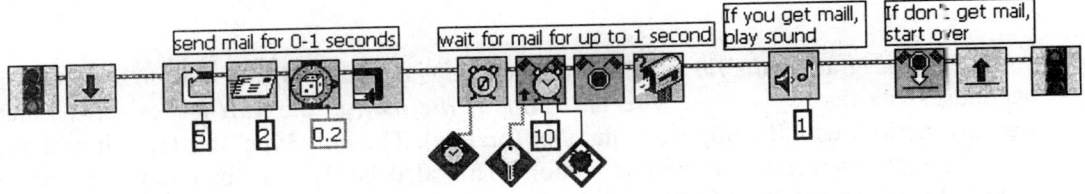

Figure 4.23. This program that spends roughly equal time listening and talking.

To make the program in Figure 4.23 useful, you would want to replace the number sent via mail to something meaningful (e.g. a container value) and the replace the sound with some other action (e.g. several mail forks).

A common mistake students make is to **send mail** in one task and **wait for mail** in another task. This never works because the **send mail** function always takes priority over the **wait for mail** function (the IR port cannot do both at once). Thus, you never end up listening. To get around this, you need to include a **wait for time** immediately after the **send mail** function. This ensures there is some time for listening.

4.4.2 Set Display

The *Set Display* function is also included in the **RCX Communication** skill badge simply because it's located on the same sub-menu, not because it involves inter-RCX communication (it can, however, be thought of as intra-RCX communication).

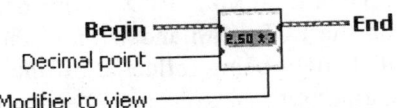

Figure 4.24. The wire terminals for the Set Display function.

The *set display* function requires 2 modifiers, one for the number to display and a second for the location of the decimal point. Any integer up to 4 digits can be displayed to the LCD. There is also a sign placeholder, which doesn't count as one of the 4 digits. This means that numbers ranging from -9999 to 9999 can be displayed. The decimal point specifies the location of the decimal point. Valid values for the decimal point modifier are from 0-3. The program below would display the sequence of numbers, 1000, 100.0, 10.00, and 1.000.

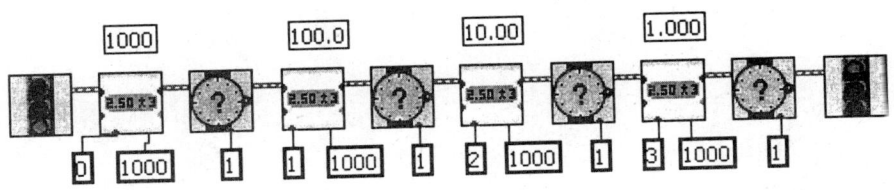

Figure 4.25. This program will display 1000, 100.0, 10.00, and 1.000.

This program *waits for mail* and then displays the *value of the mail* to the RCX for 4 seconds. Note that we cannot wire the *value of the mail* to the *wait for seconds* function because it requires a floating point number (orange). The only type of integer it will accept is the integer *numerical constant* (blue). This also holds true for other functions that require floating point modifiers (e.g. *wait for temperature*).

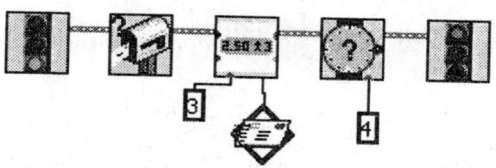

Figure 4.26. Displaying the value of the mail to the LCD.

4.4.3 RCX Communication Examples

While not strictly a communications example, this one is fun to do. The LEGO light sensor is very sensitive to infrared (IR) light along with the visible range of light. In fact, this is often the source of much frustration for our students. IR light sources are everywhere, we just can't see them!

Rather than get mad about the light sensor's overly sensitive behavior, we can try to exploit it instead. Try this: download the following program to your RCX. Mount the light sensor facing forward so the IR bean the red LED on the light sensor are sending out light in the same direction. Then hold your RCX about 6 inches from a wall. If you view the light sensor readings on the LCD with and without the program running, you will see that indeed, there is a lot of IR light being reflected off the wall. You may have to the IR power to high via the Administrator area (see chapter 2).

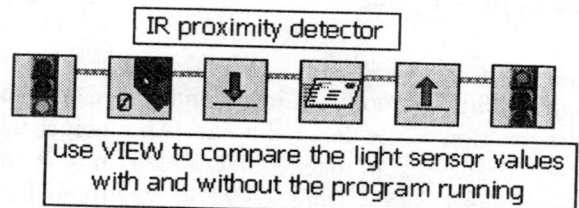

Figure 4.27. Simple program that uses the IR port as a proximity detector.

4.5 The Direct Mode Badge

Direct mode is a unique way of running programs on the RCX. Rather than downloading an entire program to the RCX as you normally do, in **direct mode** each function is sent one at a time, one after the other.

So why would you want to run in **direct mode**? The real power will be unlocked later in Chapter 5 where you will learn basic LabVIEW G-code. Then you can use your PC to perform operations too complex for the RCX. Essentially you will write programs part of which run on the RCX and part of which run on your PC. But, alas, we are not ready for that yet!

The main disadvantage of **direct mode** is that you cannot use any control structures such as loops, jumps, and task splits. If you try, you will generate an error message that indicates you can only use them in "**remote mode**" (the normal way of programming).

So why would you want to run in **direct mode** if you don't know G-code and you can't use any control structures? The answer is a bit surprising: <u>*you can run direct mode programs on top of programs that are already running!*</u>

With that said, many of the **direct mode** functions output either True or False (Boolean), which you can't really utilize unless you know some real G-code. So for now, we'll restrict ourselves to just about half of the functions: ***Begin Direct Mode***, ***End Direct Mode***, ***Wait for RCX to be in View***, ***Read Value***, ***Read and Display Value***, and ***RCX Battery Power***.

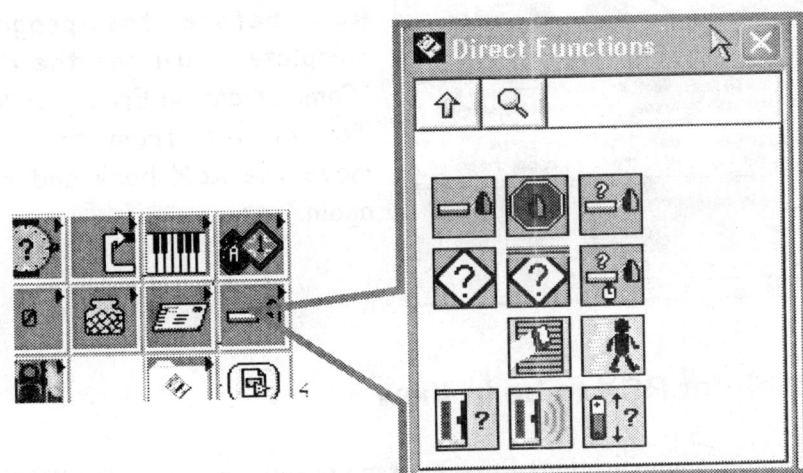

Figure 4.28. The **Direct Mode** sub-memu is accessed on the bottom right of the main **functions palette**.

4.5.1 Begin and End Direct Mode

Creating programs that run in Direct Mode is accomplished simply by replacing the standard green and red traffic lights with the ***Begin Direct Mode*** and ***End Direct Mode*** functions as shown in Figure 4.29. Just be sure to keep in mind that you cannot use any control structures (loops, forks, jumps, or task splits).

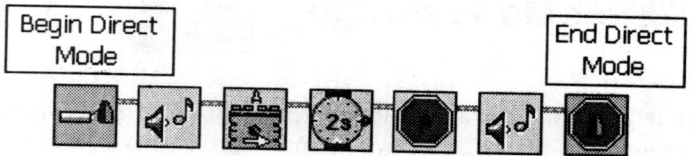

Figure 4.29. The **Begin Direct Mode** and **End Direct Mode** functions are used in place of the green and red traffic lights. Direct Mode programs execute as soon as they are received by the RCX.

In order for **Direct Mode** programs to work correctly, the RCX must remain in view of IR tower for the duration of the program. Since functions are sent one at a time, if you move the RCX before the entire program is completed, you will get a communication error. However, you can "recover" by moving the RCX back into view of the IR tower and then clicking the "try again" button on the error dialog box.

Figure 4.30. If you move the RCX before the program is complete, you'll get the dreaded "Communication Error" dialog box. To recover from this, simply move the RCX back and hit "try again."

4.1.2 Wait for RCX to be In View

You'll get the Communications Error message if you don't have the RCX in view at the start of the program too. To avoid this, you can use the **Wait for RCX to be in View** function. Just like any other **wait for** function, it will wait for the IR link to be established before proceeding. Since it doesn't hurt, most **direct mode** programs will start with this function, as shown in Figure 4.31.

Figure 4.31. The **Wait for RCX to be in view** function is very useful. Direct Mode programs won't execute until the RCX is in view of the IR tower.

4.1.3 Read Value

The *read value* function outputs any value (sensor, container, mail, timer, etc.) as a real number. In the upper program in Figure 4.32, the read value function is used to output the value of sensor port 1 to the LCD. It may seen a bit strange to have a function that does this, when you could just wire the sensor port value to the LCD directly as shown in the bottom program in Figure 4.32. However, there are many cases where a "middle man" is needed and the *read value* function serves as the middle man.

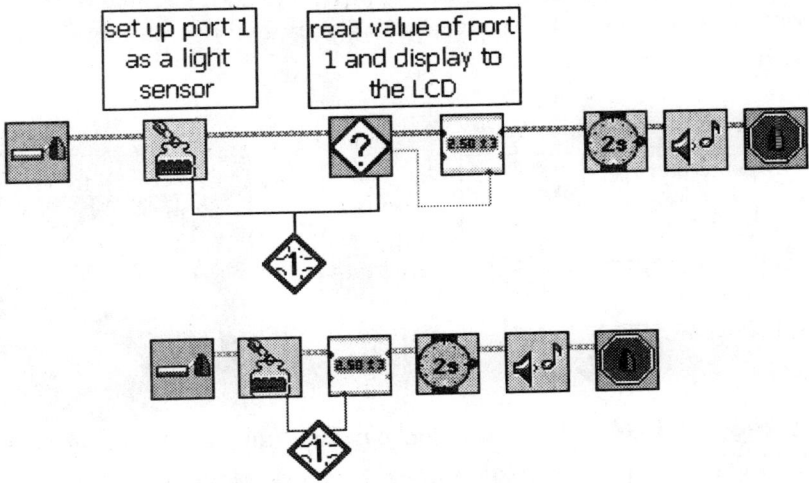

Figure 4.32. Both programs shown have the same end result. The *read value* function in the top program is used to access the values of sensor ports, container values, and mail value.

For example, in Figure 4.33 the *value of port* 1 is being used to define how long to wait (in seconds). You can't wire the *value of port 1* directly to the *wait for seconds* function because it only accepts real numbers (orange wire) and the port value only provides integers (blue wire). Try it out.

Additionally, as we will see in the next section, when we want to share sensor values with another RCX we have to use the *read value* function as a middle man, there is no way around it.

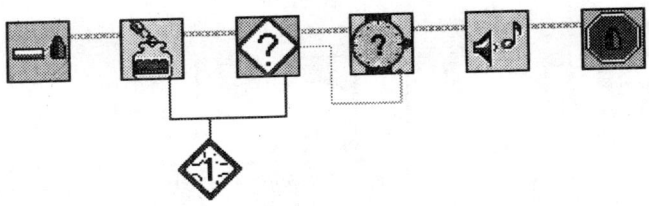

Figure 4.33. Example of a program that needs to use the *read value* function as a "middle man."

Finally, you may be wondering why we started each of the programs in Figures 4.32 and 4.33 with the *light container* function when we don't ever use containers in any of those programs. The RCX needs to know what type of sensor is connected to port 1 and by using ANY light sensor function we essentially let the RCX know that port 1 has a light sensor attached to it. Thus, we could have used any light sensor function; we just chose the *light container* function arbitrarily.

The *read and display value* function is not used as much as *read value* function, but is similar in function. Rather than outputting the value as a real number that can be used in the program, it opens a dialog box that displays the current value in real time. The program will wait until you click on the green check mark before continuing to the next function in the program.

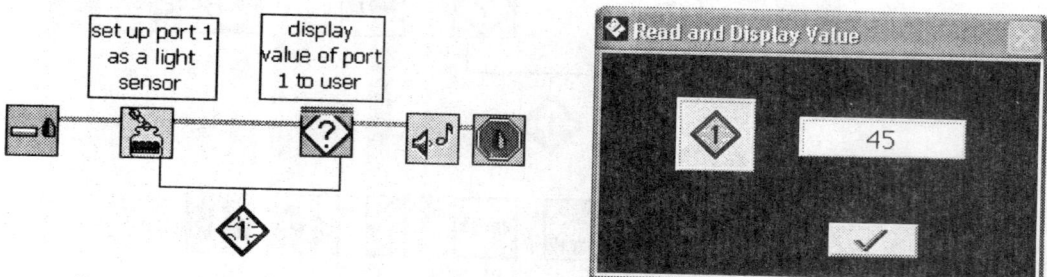

Figure 4.34. The *read and display* function allows you to see the value of a port in real time in a new window (right)

4.1.4 RCX Battery Power

The *RCX Battery Power* function returns the current battery voltage as a real number, which can be used anywhere a numeric constant is used. In the example below, we first display the battery level to the LCD (note: this is rounded down since the *Set Display* function uses an integer numeric constant) for one second. We then go on to use the battery level to control both the power level and duration of Motor A (see Advanced Output in section 4.7 for more information on controlling power levels). As the batteries wear down, the motor will run slower and for a shorter length of time; a simple kind of self-preservation.

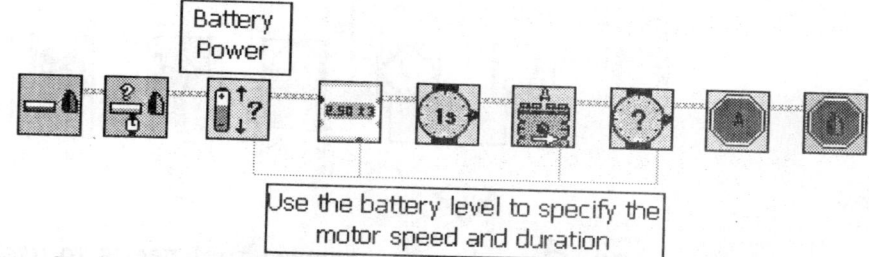

Figure 4.35. The *RCX Battery Power* functions outputs the battery voltage as a real number (0-9).

4.1.5 No Mail in Direct Mode

Since Direct Mode uses the IR communications port, it shouldn't be too surprising that you cannot use any of the mail functions in Direct Mode. If you try to run a program like the one in Figure 4.36, you will generate an error message.

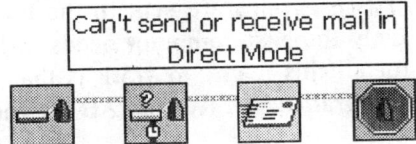

Figure 4.36. This program will generate an error message.

4.1.6 Running Direct Mode on top of Remote Mode

Finally we get to the real power of **direct mode**. As mentioned previously, the best feature of **direct mode** is the ability to run **direct mode** programs on top of **remote mode** programs. For example, downloading and running the top program in Figure 4.37 will turn on Motor A in the forward direction every 4 seconds. Running the **direct mode** program shown in bottom of Figure 4.37 would stop Motor A. But since the **remote mode** program is still running (it has an infinite loop), Motor A will start again in a few seconds.

When we covered *task splits* in Chapter 3, you learned that the RCX can mulit-task and essentially run several programs simultaneously. This works exactly the same way, except that you get to "interject" a new task via **direct mode** whenever you want.

Can you think of any instances when it would have been nice to run two programs at the same time? Now you can!

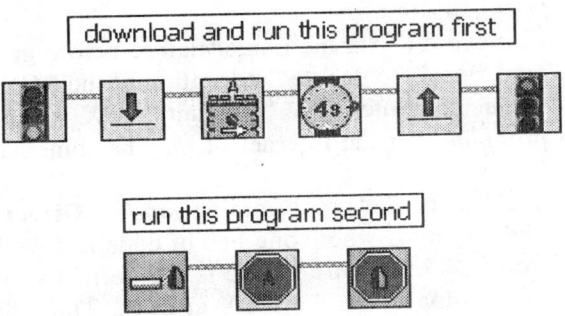

Figure 4.37. You can run two programs at the same time if one of them is a **direct mode** program.

4.6 The Internet Badge `Internet`

As the name implies, the Internet skill badge covers how to program the RCX over the Internet. The computer that receives the programs over the Internet is designated as the host computer. ROBOLAB programs are sent to the host computer via the Internet from the remote computer. Only the host computer needs to have an IR tower connected to it. The only requirement for all this magic to work is that the host computer must be running a special ROBOLAB program called ROBOLAB Internet Server.

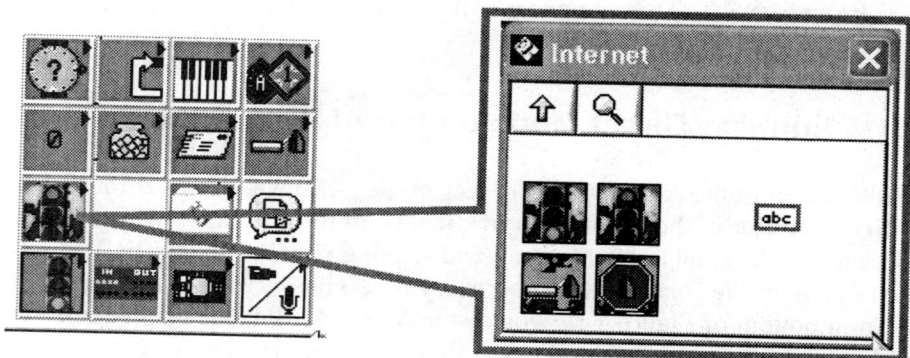

Figure 4.38. The **Internet** sub-menu is located at the bottom left of the **functions palette**.

4.6.1 Nomenclature: Remote, Direct, Internet, and Local

We have to be a bit careful with the nomenclature before getting started. We will use the terms "Internet" and "local" to denote the location of the RCX. The "local" RCX is the one right in front of your computer. The "Internet" RCX is the one at some other location which you will program over the Internet. It may be someplace very distant or simply on the next computer over.

The confusion lies in the programming *modes*: **Direct Mode** and **Remote Mode**. **Direct Mode** programming is where one line of code is downloaded to RCX at a time and was covered in section 4.5. **Remote Mode** is the normal way of programming where the entire program is downloaded to the RCX at once. Thus, there are 4 possible types of programs: local **remote mode** (this is the one you're most familiar with), local **direct mode**, Internet **remote mode**, and Internet **direct mode**. Got it?

4.6.2 Configuring the ROBOLAB Internet Server

The ROBOLAB Internet Server is a special ROBOLAB program that the host computer must run. The ROBOLAB Internet Server can be started by selecting it on the **Tools** menu as shown in Figure 4.39.

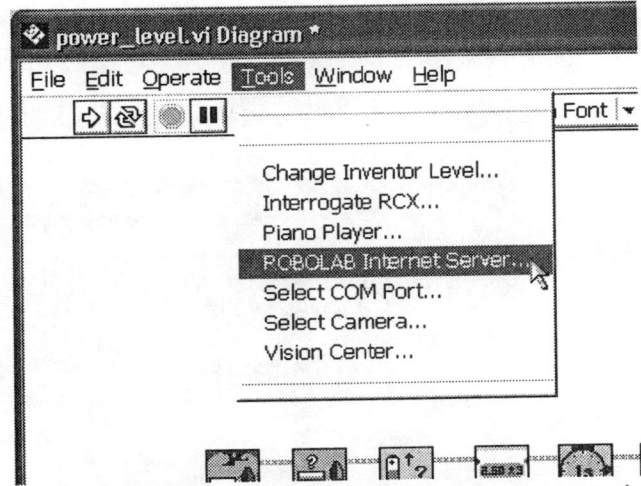

Figure 4.39. The ROBOLAB Internet Server program is located on the **Tools** menu.

NOTE: in ROBOLAB versions 2.5.0 and earlier, the **Tools** menu is called the **Project** menu as shown in Figure 4.40.

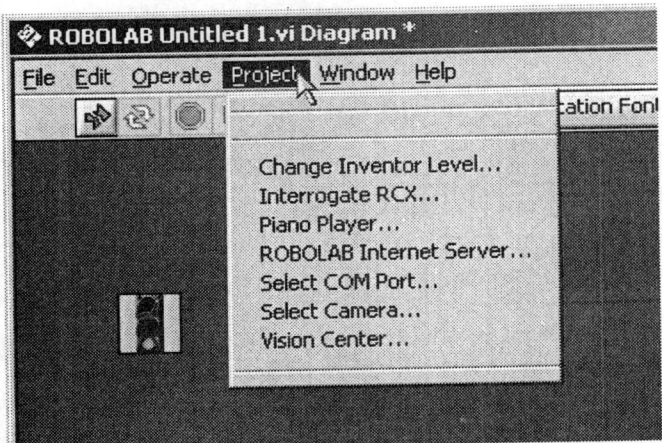

Figure 4.40. In ROBOLAB versions 2.5.0 and earlier, the **Tools** menu is called the **Project** menu.

To start The ROBOLAB Internet Server, simply select it from the **Tools** menu (Figure 4.39) and the ROBOLAB Internet Server program window will open as shown in Figure 4.41. In the main Server window there are two panes showing the **Machine Access** list and the **Exported VI** list. At the bottom of the window the host computer's IP address is displayed.

The **machine access** lists controls which computers are allowed to send ROBOLAB programs to the host computer. The default list includes the *localdomain* and an asterisk (*). The *localdomain* allows all computers on the local network access. The asterisk (*) allows all computers access (easy to setup, but very unsafe from a network security standpoint). You can change the list of computers by simply entering either the computer's name or IP address as shown in Figure 4.42. In the example shown, only the 2 computers specified would be allowed to send ROBOLAB programs to the Host computer.

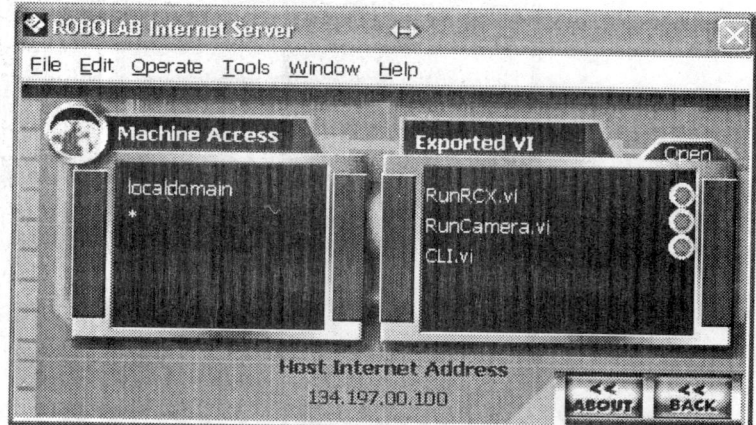

Figure 4.41. The host computer needs to be running the ROBOLAB Internet Server program.

Figure 4.42. ROBOLAB Internet Server with the **Machine Access** limited to two specific computers.

In case you haven't noticed it by now, all your Inventor mode programs have the *.vi* filename extension and your Investigator programs have the *.llb* filename extension. A library (*.llb*) is simply a collection of *.vi* files (the *vi* stands for Virtual Instrument). With this in mind, the Exported VI list, shown in the right hand pane of Figure 4.41, should make some sense. The list simply specifies which Virtual Instruments (VI's) can be run over Roboserver. The *RunRCX.vi* is used to run standard ROBOLAB programs. The *RunCamera.vi* is used to run Vision Center programs. Finally, the *CLI.vi* is used to run a Control Lab Interface (CLI) program (see Appendix B for more information on the CLI). You can add your own list of VI's to this list, **BUT WE DON'T RECOMMEND EDITING THE LIST OF VI's**! If you are interested, you can click on the green circle and open the corresponding VI and examine it. You will see that they were written in LabVIEW G-Code, not ROBOLAB. For more about LabVIEW, we highly recommend getting a book such as *LabVIEW for Everyone* or the *Student Edition of LabVIEW*, both available from Prentice-Hall.

When a program is submitted over the Internet, the **Exported VI** list will briefly highlight the VI being executed. For 99% of the time, this will be the *RunRCX.vi* as shown in Figure 4.43.

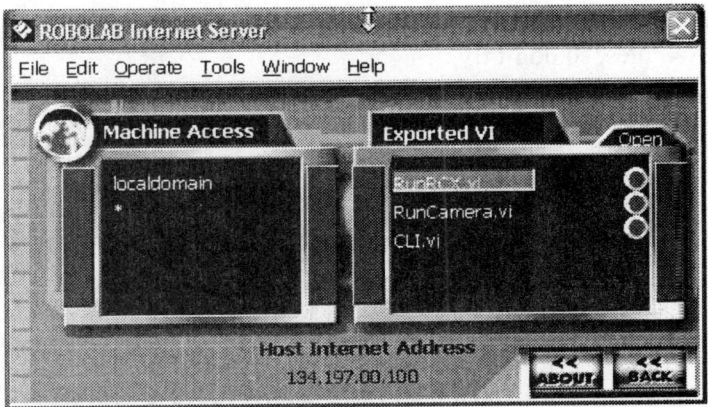

Figure 4.43. *RunRCX.vi* will be highlighted in red in the ROBOLAB Internet Server window as the program is received on the host computer.

Note that in ROBOLAB versions 2.5.0 and earlier, the Internet Server window appears slightly different (as shown below). It does not show the Host computer's IP address. This means that you will have to determine the IP address by examining the network properties or asking your local friendly Systems Administrator.

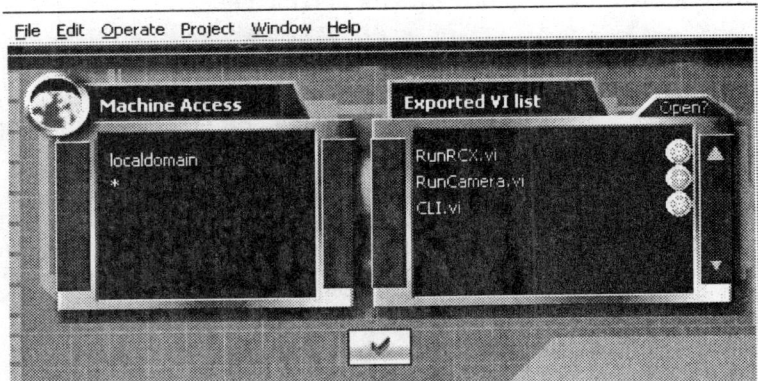

Figure 4.44. In ROBOLAB version 2.5.0 and earlier, the ROBOLAB Internet Server window does not show the Host IP address. Clicking the green check mark closes the Internet Server window.

4.6.3 Internet Begin and End

Submitting programs over the Internet is not much different than writing a normal ROBOLAB program (which you have ample experience with by now). The only differences are that the ***Internet Begin*** and ***Internet End*** functions are used in place of the normal ***Begin*** and ***End*** traffic lights as shown in Figure 4.45. The host computer's IP address must be wired to the Internet Begin function using the pink ***string constant*** (Figure 4.39). The host computer's IP address is always shown at the bottom of the ROBOLAB

Internet Server window (Figure 4.41). Note that we've made up a fictitious IP address for these examples, so don't try using it!

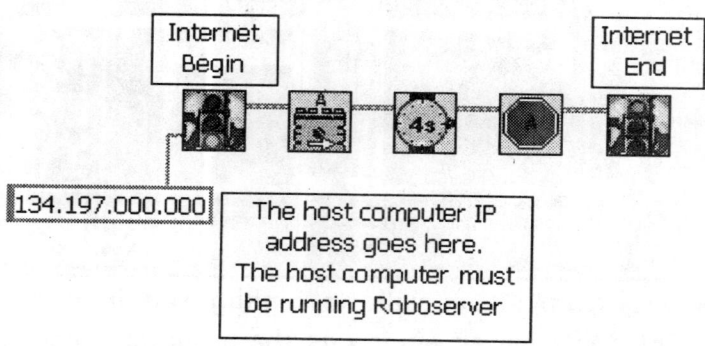

Figure 4.45. *Internet Begin* and *Internet End* functions replace the usual *Begin* and *End* traffic lights. The host computer IP address must also be wired to the Internet Begin function

For many local networks, using the host computer's Network ID (a.k.a. the computer name) will also suffice as shown in Figure 4.46. This is often more convenient than using the IP address, which is sometimes difficult to remember.

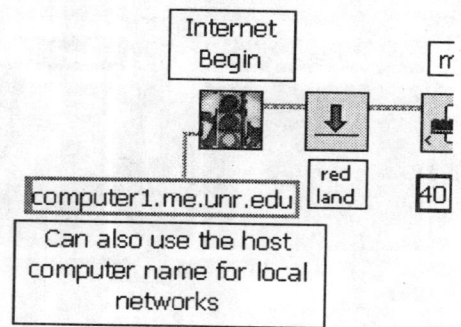

Figure 4.46. The computer name can be used in place of the IP address on most local networks.

Figure 4.47 shows the infamous LEGO cockroach from Chapter 3, this time redone as an Internet program. Clicking the white run arrow would send this program over the internet to the host computer specified. Now you can create infestations anywhere on the planet!

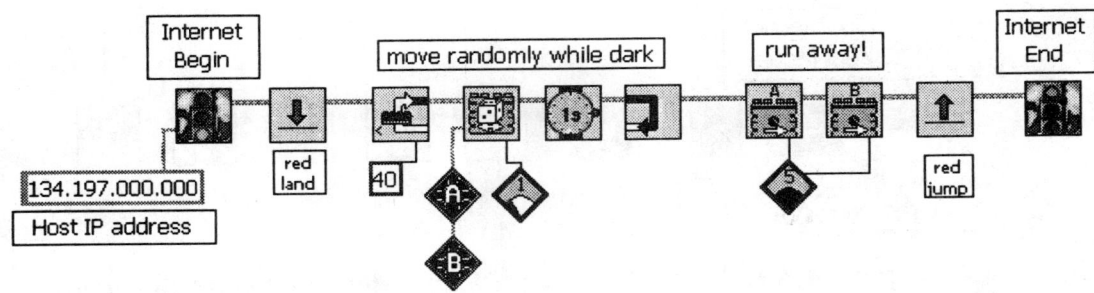

Figure 4.47. The cockroach program from Chapter 3, redone using the Internet.

As soon as you "run" an Internet program, the *ShowNet.vi* status window will open on the local computer (usually <u>very</u> briefly, depending on the connection speed) as shown in Figure 4.48.

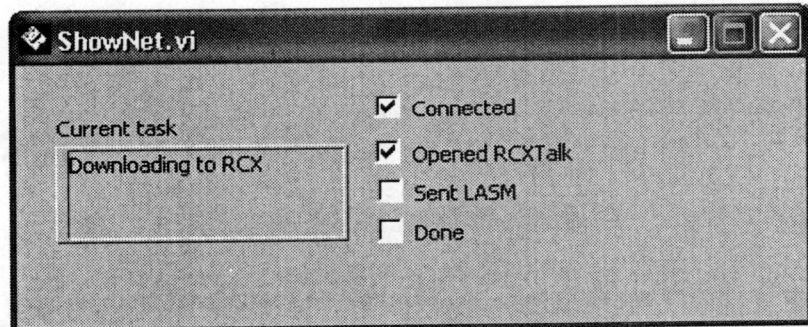

Figure 4.48. The status window will indicate progress on the remote computer.

4.6.4 Internet Direct Mode

Internet Direct Mode combines both **Internet Mode** and **Direct Mode** and allows you to directly run programs over the Internet on a remote RCX. Just as before, the Internet RCX must be in IR communication with a computer running ROBOLAB Internet Server.

In the figure below, we've taken the **Direct Mode** program from Figure 4.35 and redone it using **Internet Direct Mode**. In **Internet Direct Mode**, you have the same programming limitations as **Direct Mode**; you cannot use many functions including control structures (loops, jumps, and forks) and some *wait for* function (e.g. *wait for container*).

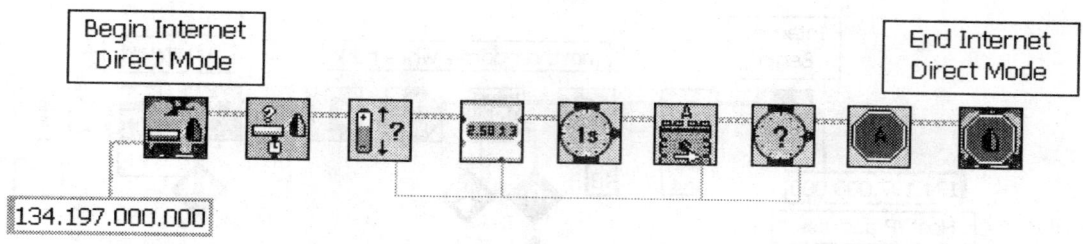

Figure 4.49. Using **Internet Direct Mode** you can run programs in **Direct Mode** over the Internet.

4.6.5 Controlling Two RCX's Simultaneously

One of the neatest tricks that we've stumbled upon is the ability to run two **remote mode** programs at the same time! This works as long as one of the programs is an Internet program. In Figure 4.50, we've created two programs, each with their own green and red traffic lights. Normally, this wouldn't work (only one program would be downloaded to the RCX). However, since the upper program is being sent to an Internet RCX and the lower program is being downloaded to the local RCX, it all works!

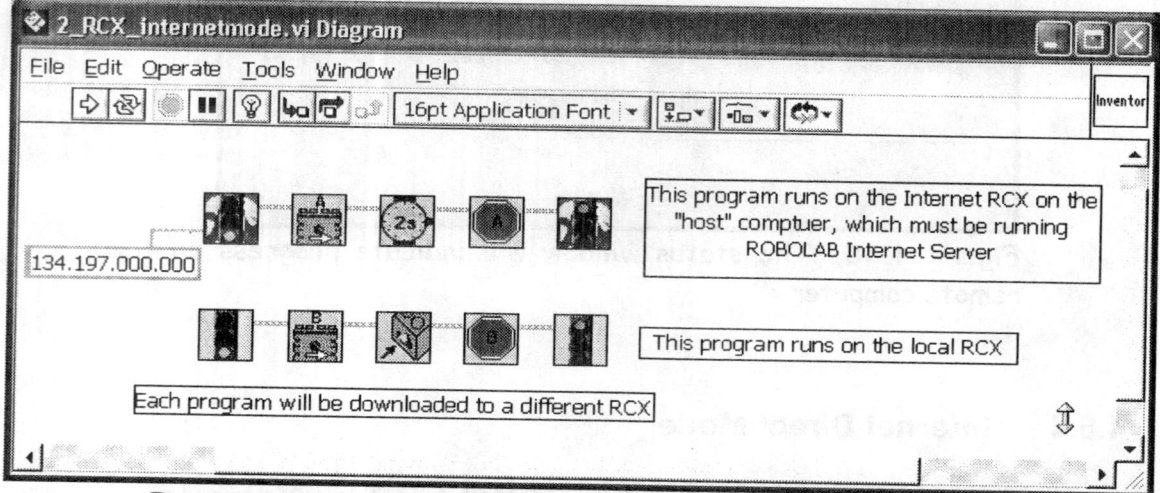

Figure 4.50. Using **Internet Direct Mode** you can run programs in **Direct Mode** over the Internet.

Okay, this is pretty cool trick, but it gets more interesting. We also have the ability to share data (sensor, container, or mail) between RCX's if we run one of the RCX's in **direct mode** rather than **remote mode**. Then the other RCX can access its data as shown in Figure 4.51. Here the local RCX's program uses the value of the Internet RCX's light sensor.

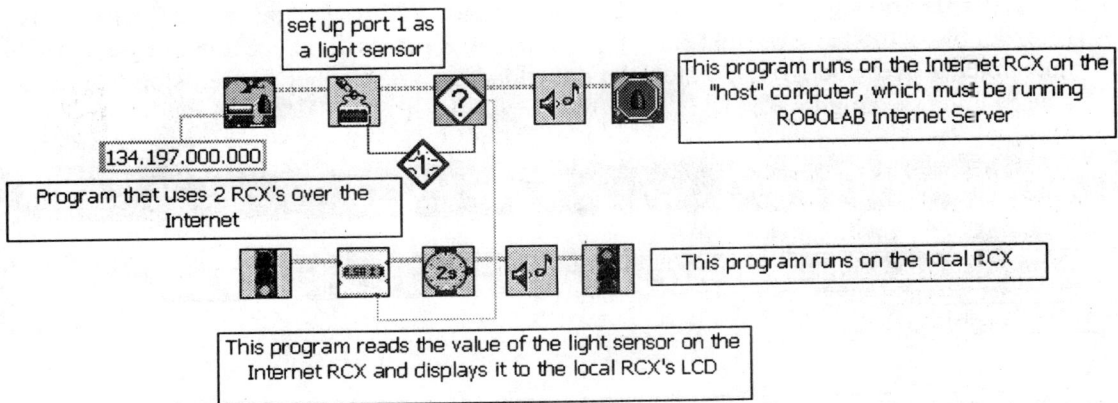

Figure 4.51. Using *Internet Direct Mode* you can share sensor data over the Internet.

The quirky nature of the *read value* function was mentioned previously in section 4.5.3. In Figure 4.52, we didn't use the *read value* function so the local red container is filled with the local value of input port 1, which is not what we intended. Looking at Figure 4.52 and thinking from your computer's perspective, isn't at all obvious which RCX the *value of port 1* function refers to; the local RCX or the Internet RCX. To avoid this ambiguity, the *read value* function was created to act as a "middle man." In Figure 4.51 it is clear which RCX's port value is being displayed.

The only other tricky point is that the sensor data from the Internet RCX will only be read once - when the local RCX's program is downloaded. In the example shown, running the local RCX's program over again will not reflect any changes in the light sensor values.

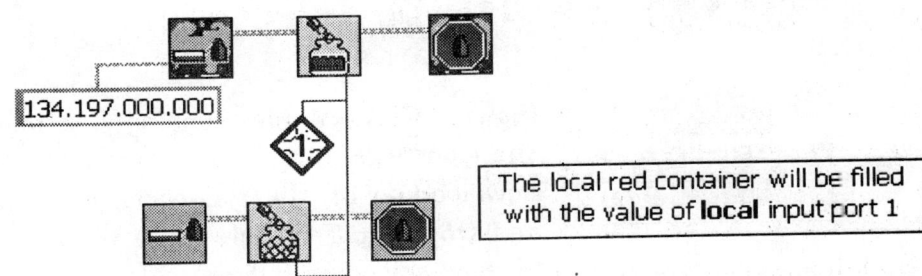

Figure 4.52. This does not produce the same result as Figure 4.51. The local container is filled with the local sensor port value. You need to use the *read value* function to share data.

We could have just as easily switched the roles of the local and Internet RCX's with the Internet RCX using the local RCX's data, but that would still lead to the sensor data only being shared when the Internet program is downloaded.

To get around this limitation, we can run both RCX's *continuously* in **Direct Mode**. Then the data sharing can go both ways in real time. Running the program shown in Figure 4.53 will put the light sensor data of the Internet RCX into the red container of the local

RCX. As mentioned back in section 2.6 (see Figure 2.23), selecting **Run Continuously** (the double white arrow) instead of the normal **Run**, will cause the program to be downloaded over and over. To stop the downloading, press either the red stop sign or the pause button (see Figure 4.54).

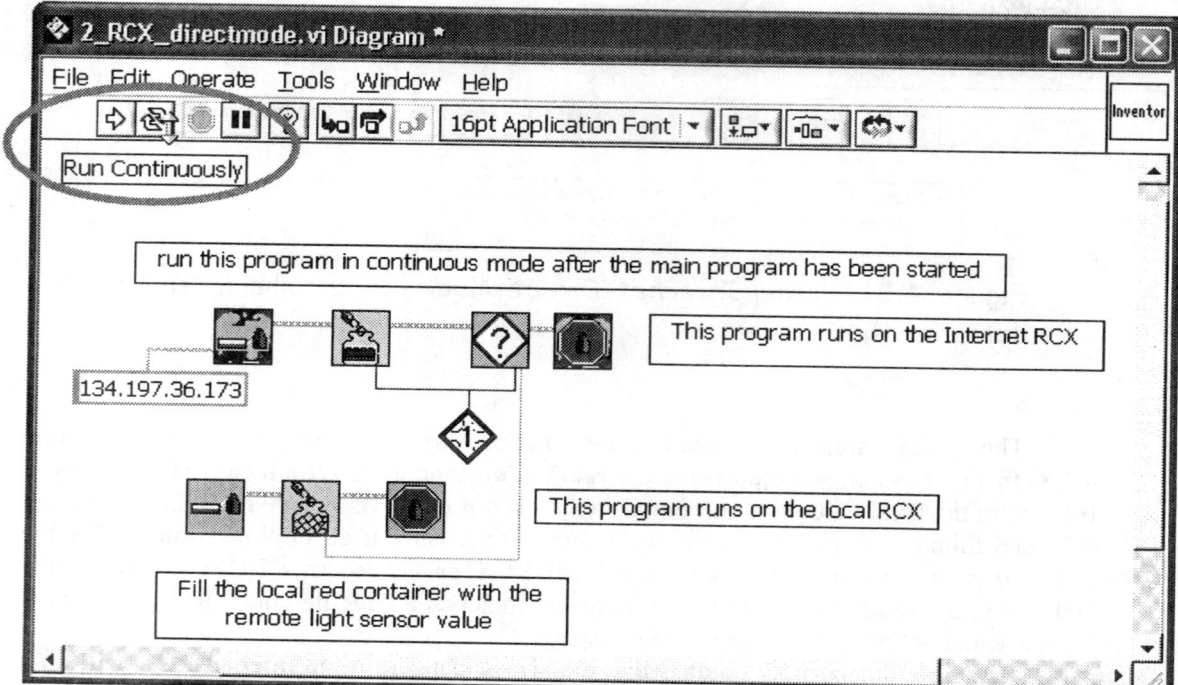

Figure 4.53. Using *Direct Mode* for both RCX's lets you can share data between RCX's in real time. Running Continuously will cause the program to repeat.

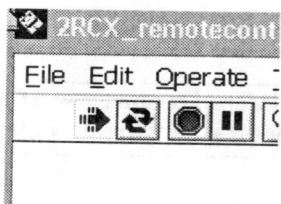

Figure 4.54. While Running Continuously, the icons will change. To abort the repeated downloading of the programs, click on the red stop sign. To simply pause the execution, click on the pause button.

So now we know how to pass data back and forth between RCX's. But what good is it if you can't use any control structures? Recall, that **direct mode** programs can be run on top of **remote mode** (normal) programs. Thus, we can first download and run a program like the one shown in Figure 4.55. The program is a generic program for steering a car using two containers. Each motor will run either in the forward or reverse direction, depending on the value in the blue and red containers.

By running the two direct mode programs shown in Figure 4.56 on top of the other program (which should be in program slot #3), your buddy can drive your car over the Internet using his/her touch sensors. Touch sensor 1 will control the direction of Motor A and touch sensor 2 will control Motor B. Of course, realistically the delay caused by the

Internet will make it hard to do any precision driving. To maintain constant communication between the RCX and IR tower, we've often directly mounted the IR tower on the RCX as shown in Figure 4.57.

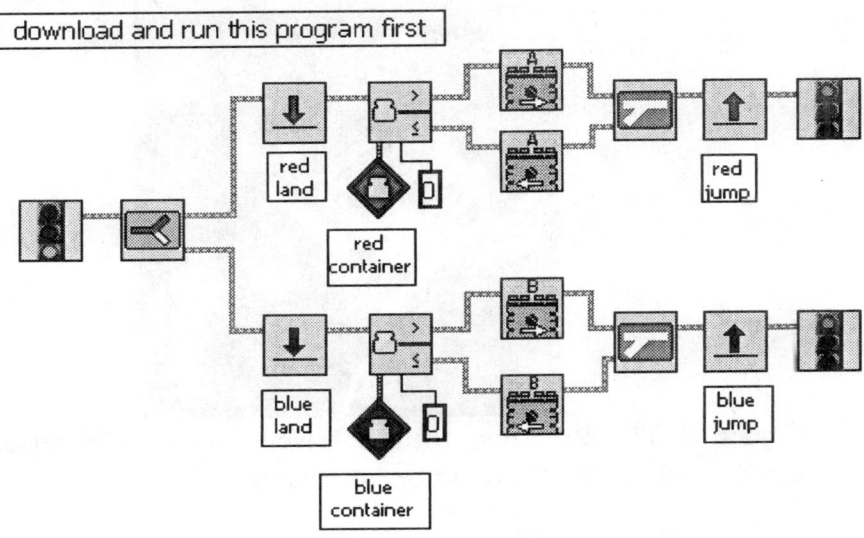

Figure 4.55. A simple program to steer a car. Download this to program slot #3.

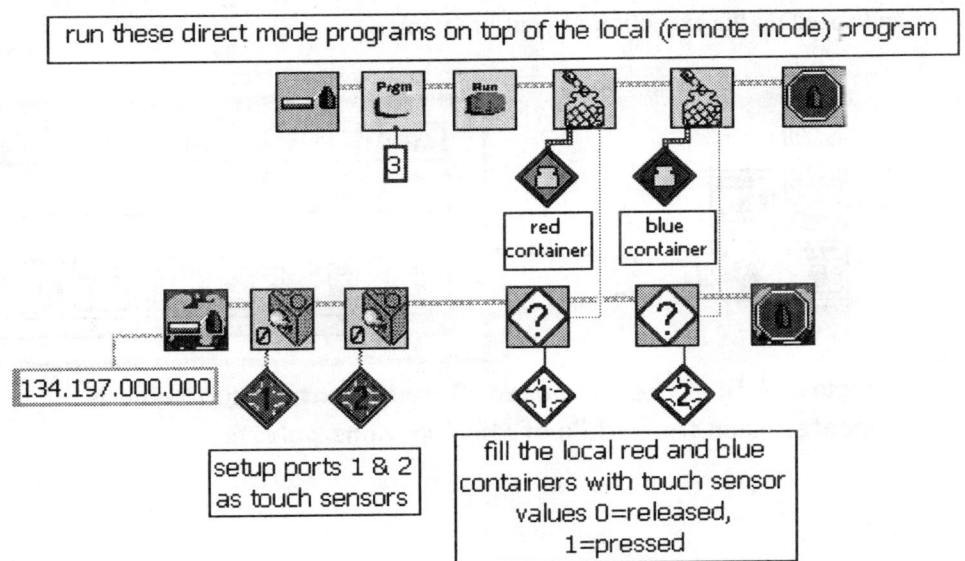

Figure 4.56. Run these two **direct mode** programs continuously and your buddy can drive your car using 2 touch sensors.

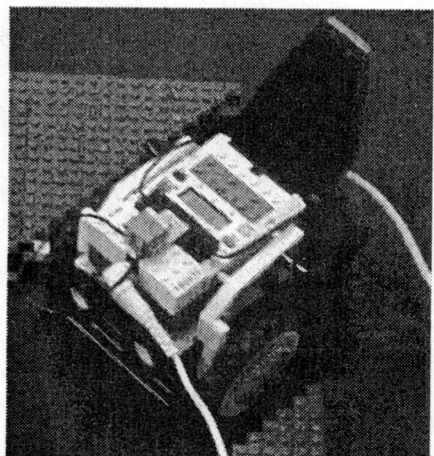

Figure 4.57. Mounting the IR tower to the RCX ensures the direct mode programs won't be interrupted.

4.7 The Advanced Output Badge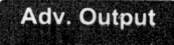

By now you've gotten pretty used to using the LEGO motors. The Advanced Output skill badge will provide you with a few more tools to help you both refine motor control and speed up the execution of your programs.

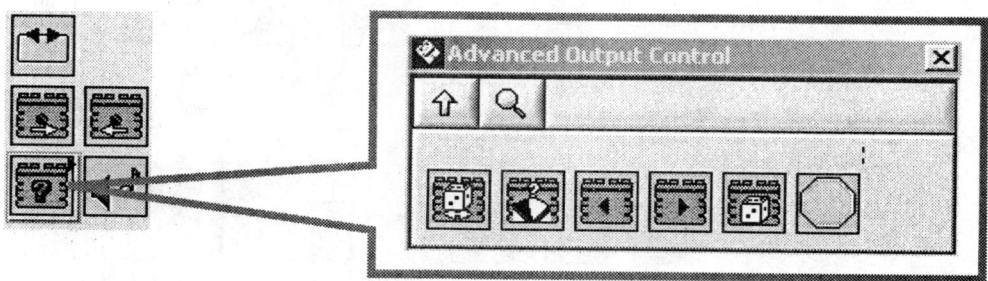

Figure 4.58. The Advanced Output Control sub-menu is located near the middle of the **functions palette**.

4.7.1 Advanced Output sub-menu

It turns out the simple turn *Motor A forward* function is actually a combination of three individual functions, each of which can be accessed via the Advanced Output Control sub-menu. To turn Motor A on in the forward direction we actually *set the direction* to forward, set the *motor power level* to full, and then *turn the motor on*.

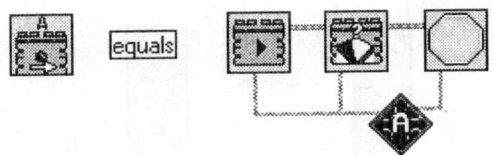

Figure 4.59. The *Motor A forward* is actually a combination of 3 basic commands.

Why go to all this trouble? The short answer is: to make faster executing programs. Why execute three commands to change the motor direction when you can accomplish it with a single command?

We also have access to some motor controls we didn't have before. For Example, we can now set the direction randomly if we so desire.

4.7.2 Pulse Width Modulation

The RCX varies the motor power using a technique called **Pulse Width Modulation**. To turn the motor on at full power we simply apply 9 Volts to the motor continuously. To turn the motor on at 50% power, we quickly switch the Voltage on and off. Figure 4.60 shows a series of pulses being sent to the motor that represents a 50% duty cycle, meaning the Voltage is only on for 50% of the time. Since we are supplying Voltage to the motor only 50% of the time, the motor runs at half power. We can further decrease the motor power by decreasing the width of the pulses (thus, the term pulse width modulation). Figure 4.61 shows a 25% duty cycle.

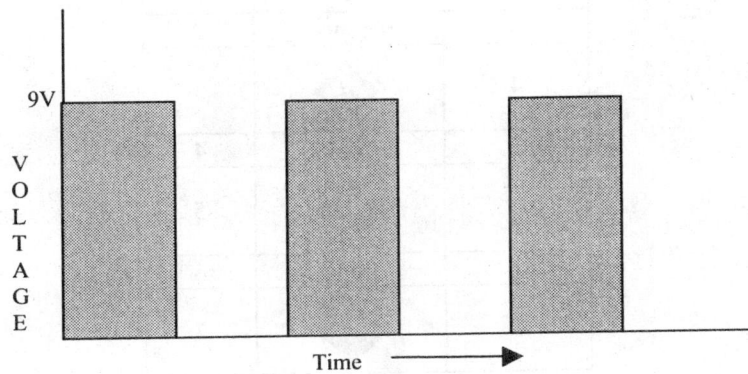

Figure 4.60. Pulse Width Modulation with 50% duty cycle.

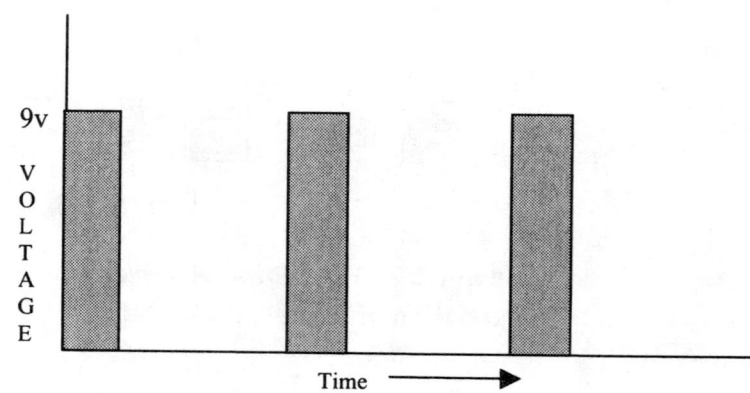

Figure 4.61. Pulse Width Modulation with 25% duty cycle.

4.7.3 Controlling Motor Power

Okay, so now you know how the RCX controls the motor power with pulse width modulation. Up to this point we've controlled the power level of the motors by simply using one of the 5 power level modifiers. However, the LEGO motors actually can be run at 8 different power levels with each level corresponding to roughly 12.5% increments in duty cycle.

RCX Power Level	Power Level Modifier	Integer Value
1	◇1◇	0
2	◇2◇	1
3	none	2
4	◇3◇	3
5	none	4
6	◇4◇	5
7	none	6
8	◇5◇	7

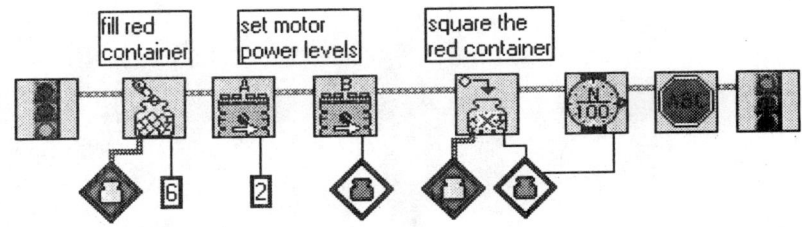

Figure 4.62. Setting the motor power using integers and container values. The motors will run for a total of 0.36 seconds in this program.

4.7.4 Power versus Speed

It's easy to confuse power and speed. We usually want to control motor speed to make a LEGO car go faster or slower, but we actually are controlling the motor power. The missing links between power and speed are **mass** and **gear ratio**.

A very heavy vehicle will take lots of power to get moving. Thus, a low power level will appear to produce a low motor speed. Likewise, if we use a big gear ratio (a big gear driving a little gear) then controlling motor power will also appear to control motor speed. This is because a big gear ratio makes the vehicle appear to have a lot of inertia. Think of trying to get your car moving in from a dead stop using 4th gear or trying to pedal a bicycle while using a big gear ratio – it's hard! In these cases, varying motor power will seem to control motor speed.

The confusion arises when you have either a very light vehicle (which is pretty typical for LEGO vehicles) or a low gear ratio (small gear on the motor turning a big gear on the wheel). In these two cases, the motor has more than enough power to get the vehicle moving. Even at the lowest power level the vehicle will take off a top speed. Changing power levels will appear to have no affect on the vehicle speed!

Thus, the bottom line is to remember that you are actually controlling motor power (using pulse width modulation) not speed. However, by varying the vehicle mass and/or gear ratio you can use power to affect speed.